Cultural Economies of Locative Media

OXFORD STUDIES IN MOBILE COMMUNICATION

The Oxford University Press series on Studies in Mobile Communication focuses on the social consequences of mobile communication in society. The series includes forthcoming work on, for example, mobile communication in developing countries, location based services, among blue-collar workers in the United States, and parent/child communication.

Series Editors
Rich Ling, *Nanyang Technological University, Singapore*
Gerard Goggin, *University of Sydney, Australia*
Leopoldina Fortunati, *Università di Udine, Italy*

Haunting Hands: Mobile Media Practices and Loss
Kathleen M. Cumiskey and Larissa Hjorth

*A Village Goes Mobile: Telephony, Mediation,
and Social Change in Rural India*
Sirpa Tenhunen

Negotiating Control: Organizations and Mobile Communication
Keri Stephens

CULTURAL ECONOMIES
OF LOCATIVE MEDIA

Rowan Wilken

OXFORD
UNIVERSITY PRESS

OXFORD
UNIVERSITY PRESS

Oxford University Press is a department of the University of Oxford. It furthers
the University's objective of excellence in research, scholarship, and education
by publishing worldwide. Oxford is a registered trade mark of Oxford University
Press in the UK and certain other countries.

Published in the United States of America by Oxford University Press
198 Madison Avenue, New York, NY 10016, United States of America.

Library of Congress Cataloging-in-Publication Data
Names: Wilken, Rowan, author.
Title: Cultural economies of locative media / Rowan Wilken.
Description: New York, NY : Oxford University Press, [2019] |
Includes bibliographical references and index.
Identifiers: LCCN 2019012352 | ISBN 9780190234911 (hardback) |
ISBN 9780190234928 (pbk.) | ISBN 9780190234935 (updf) |
ISBN 9780190234942 (oso) | ISBN 9780190070632 (epub)
Subjects: LCSH: Location-based services—Social aspects. |
Location-based services—Economic aspects.
Classification: LCC HM851 .W55154 2019 | DDC 302.23/1—dc23
LC record available at https://lccn.loc.gov/2019012352

9 8 7 6 5 4 3 2 1

Paperback printed by Marquis, Canada
Hardback printed by Bridgeport National Bindery, Inc., United States of America

In memory of Peter Bayliss and Scott Ewing

CONTENTS

FIGURES

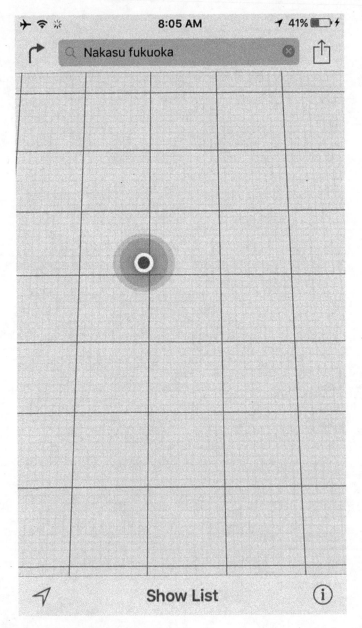

Figure 0.1. Apple Maps screenshot.
Photograph: author.

PREFACE

[A] presence, a visible presence, is sometimes most eloquently conveyed by a disappearance.

—Berger (2001, 246)

Maps reveal and they conceal. Sometimes, as is the case with the Apple Maps screenshot appearing in Figure 0.1, they are most revealing for what they conceal, as well as for what they fail to reveal (Bonnett 2014; Monmonier 1996; D. Wood 1992). But how might we get at that which is concealed or yet to be revealed? Elsewhere I have argued that description offers a productive means for understanding familiar, routinized technology use (Wilken 2017). Description unsettles the familiar; it provides us with a means of engaging with the inconspicuous and the hidden "as an object of attention" (Clucas 2000, 25).

What, then, is the scene of the image? It is June 9, 2016 (I know this from my iPhone photo library metadata), and I am in the Daimyo district of Fukuoka, Japan. As an international traveler without roaming data activated, I have put my phone into "flight mode" and I am relying on Wi-Fi for mobile internet access, as indicated in the top left of the screen image by the presence of the familiar Wi-Fi symbol of a dot with curved lines radiating from it. Unbeknownst to me, a month prior to my visit it became legal for a "non-Japanese foreigner" to use his or her own phone in Japan for up to 90 days at a time, so long as the phone carried an FCC or a CE (if not a Japanese Giteki) mark as well as a Wi-Fi Alliance mark (Yoshida 2014b, 2016). Prior to this, regulatory complexities specific to Japan— including the long-standing "Mobile Phone Improper Use Prevention Act (携帯電話不正利用防止法)"—meant that international travelers had to rely on rented handsets, or, more commonly, rented Wi-Fi "eggs" (portable Wi-Fi hotspot devices) (Yoshida 2014a). Unaware of this law change, I was using a Wi-Fi "egg" for internet access.

Besides grid lines and a large blue dot, what else does this image reveal? At bottom left in this version of Apple Maps is a swallow-tailed directional arrow which, when tapped, activates "compass mode." This makes a small blue triangle appear atop the blue locational dot, which swivels as I move my phone so that the tip of the triangle always points north. And at the bottom right is the "i" symbol; tapping this leads to Apple Maps settings, as well as to a list of the firms that Apple has data licensing arrangements with that support the provision of its iOS maps. Having forgotten to charge my phone the night before (the battery is already down to 41% at 8:05 a.m.), the lightning bolt symbol at top right reveals that I am using the device while plugged into a Japan-specific iPhone plug (bought the previous day from a local ¥100 store). To the left of the power indicator, the smaller stylized swallow-tailed directional north arrow reveals that an app—Apple Maps—is accessing my location.

Before heading out to spend the day attending an international communication conference—which is the ostensible reason for my visit to this part of Japan—I have opened Apple Maps and typed my search term, "Nakasu fukuoka," into the search box—an attempt, on my part, to orient myself prior to a planned catch-up with colleagues in a different part of the city later that evening. While the spinning wheel at top left suggests that some network activity is taking place, my requested toponym search has not yet loaded. I am left waiting (Farman 2018a). All that is visible is a large blue-and-white circle atop a graticule (the network of lines representing meridians and parallels over which the map is usually represented). My "position" is marked by a large dot floating mysteriously in a sea of nothing.

Struck by how disconcerting I find this "glitch" to be—John Berger (2016, 222) once remarked that "people need straightaway to pinpoint where they are [. . .] as if they are pursued by doubts suggesting that they may be nowhere"—and bemused by the fact that I am at a conference to present on mobile location-based services and my own phone is struggling to locate me, I capture this moment via screenshot image.

But, what lays behind this scene? What is there in this image that cannot be seen? To begin to answer these questions, it is important to consider the location capacities of an Apple smartphone. Each iPhone comes with a GPS (global positioning system) chip inside it. GPS is a constellation of satellites and support infrastructures established and maintained by the US Department of Defense (Ceruzzi 2018; Sturdevant 2012). A GPS device determines its position by trilateration with at least three (but preferably four) of a possible 31 satellite signals (Zahradnik 2017b). There are two issues with this GPS triangulation process that affect smartphone users. The first issue is the significant elapsed time it takes for a

GPS receiver to determine fully its location, which is known as time-to-first-fix (TTFF). The second issue is that GPS signals have difficulty "penetrating buildings . . . and canyons, including urban skyscraper canyons" (Fleishman 2011).

With respect to the first issue, TTFF is determined in the following way. Each GPS satellite "broadcasts a time stamp, its current location and some less precise location information for other GPS satellites" every 30 seconds (Fleishman 2011). In order to obtain the full list of satellite locations, 25 of these broadcasts are required—a process that takes 12.5 minutes (Fleishman 2011). However, "if you know the position of four satellites and the time at which each sent their position information," your GPS receiver is able to "calculate to within 10 meters the latitude, longitude, and elevation of your current location along with the exact current time" (Fleishman 2011). With one less satellite, elevation is lost, "but a device can still track movement fairly accurately" (Fleishman 2011).

With respect to the second issue, in order to speed up the TTFF process and address urban access challenges, each iPhone draws on cell tower and Wi-Fi signals to access what is called Assisted GPS (AGPS). Whenever a GPS signal is weak or impeded in some way, smartphones will consult with AGPS servers. These servers store information about the locations of satellites. And, because these servers are accessed via the cellular network, satellite access is not required in order to obtain satellite location information (Zahradnik 2017a). Herein lies the benefits of AGPS. Rather than having to rely on "live downloads of position data from satellites," satellite trajectories can be estimated accurately enough to determine the likely position of satellites (Fleishman 2011). These estimates can be used to obtain a fix, and, once this fix is achieved, more accurate and up-to-date positional data are obtained (Fleishman 2011). Positional estimates of this sort can be downloaded via a network connection almost instantaneously, "or even calculated right on a device" (Fleishman 2011). Further clues are provided by consulting the current time: "With a precise current time, fragmentary satellite data can be decoded to gain a faster lock or figure out the appropriate information to use" (Fleishman 2011). AGPS is thus important in that it ameliorates some of the delay and urban access issues associated with calculating TTFF.

In addition to the preceding, efforts have also been made by Apple to hasten the TTFF process in such situations where a device is solely reliant on Wi-Fi to fix its position. How they are said to have done this is by "caching subsets of data about nearby networks and towers to reduce network activity and speed up such lookups" (Fleishman 2011). This is also done in order to shift "some computation to the phone or tablet and away from location services" (Fleishman 2011) in an effort to further speed up the geolocation process.

With the appearance of the large blue dot, what I was witnessing was the first stage of the Apple Maps app's attempt to locate me: my phone consulting the local database. The dot becomes reduced in size as further positional information is obtained, drawing on cellular network and Wi-Fi sources, in order to calculate a more precise trilateration (Fleishman 2011). As Glenn Fleishman (2011) explains, "this data is also used to provide more clues into decoding the best GPS satellite information, allowing the use of quite small fragments of data or even raw signals to get a better lock." Eventually, the once large dot becomes a single smaller dot "when iOS is confident it has a solid GPS lock" (Fleishman 2011).

The fact that my blue circle failed to reduce suggested something was awry. And, indeed, for users of Apple Maps, AGPS is not without its own technical limitations. As Steven Fisher (2011) explains, the iPhone, as shipped by Apple, "is completely dependent on the Internet for map tile data." What this means is that, "without a data signal, whether WiFi or 3G, the iPhone is unable to show you a map" (Fisher 2011). This is because the iPhone is attempting to extract a graphical map from its cloud servers, and while "the image the cloud returns can be beautifully rendered and completely up-to-date," without cloud access the iPhone is unable to load any form of properly rendered map (Fisher 2011). What I was experiencing when I took this image, then, was likely some kind of temporary network failure preventing the iPhone from fixing my position with precision, and preventing Apple Maps from downloading a graphical map from Apple's cloud servers. Had I been able to load data on this particular location within Apple Maps prior to this network problem (such as by searching for it the night before), this information would still have been viewable while offline (Sande 2012).

I have prefaced this book with discussion of this screenshot image (Figure 0.1), as it acts as a useful scene-setter for the discussion and concerns that follow. It also serves to illustrate that the simple act of determining one's location via smartphone applications, such as Apple Maps, is far from trivial. While GPS "is, in fact, rocket science at many levels" (Fleishman 2011), other forms of location determination carry their own technical and infrastructural complications, and each are remarkable in their own ways. Also remarkable is the fact that we rarely notice them unless they fail. A key reason for this is that we have been quick to integrate location-based services, especially those associated with smartphone use, within the routines and rhythms of our day-to-day lives. Indeed, so rapid and so thorough has this integration been that we often only notice their operation during moments of failure or delay. This familiarity, however, belies many complicated infrastructural and business arrangements that make these services possible; it belies the complexities, contradictions, and rich messiness of our everyday usage of these devices; and it belies the manifold legal, regulatory, and policy-related issues that are associated with business, government, and end-use of mobile location-based services. These are among a core suite of concerns that this book aims to explore in the chapters that follow.

ACKNOWLEDGMENTS

This book took much longer to complete than anticipated, and finishing it would not have been possible without the encouragement, assistance, and support of the following people, all of whom have helped at different points along the way and in a variety of capacities: Kath Albury, Michael Arnold, Carlos Barreneche, Peter Bayliss, Andrew Bednarz, Jean Burgess, Teresa Calabria, Angela Daly, Matt Duckworth, Scott Ewing, Jason Farman, Eve Forrest, Jordan Frith, Neil Gardiner, Martin Gibbs, Sandy Gifford, Jock Given, Ravi Glasser-Vora, Gerard Goggin, Larissa Hjorth, Heather Horst, Sharryn Knight, Grace Lee, Christian Licoppe, Ben Light, Ramon Lobato, Chris Marmo, James Meese, Esther Milne, Bjorn Nansen, John and Shirley Olsen, Erika Polson, Kane Race, Ellie Rennie, Ingrid Richardson, Brady Robards, Raz Schwartz, Emily van der Nagel, Andrew Vincent, and Hannah Withers.

I wish to express my gratitude to Hallie Stebbins, and to Sarah Humphreville at Oxford University Press, for their help in realizing this book and for their patience; the Studies in Mobile Communication series editors, Gerard Goggin, Rich Ling, and Leopoldina Fortunati, for supporting this book; and all those who gave generously of their time and shared their insights through interviews and focus groups for the project that informed this book.

I also wish to thank the artists and organizations who granted me permission to reproduce the images included in this book. Every effort has been made to trace and pay all the copyright holders.

Special thanks go to four people. To Julian Thomas, for his faith in this project—from its very earliest stages as a draft grant application, through to its completion as this book—and for his patience, ongoing encouragement, and continued support. To Lee Humphreys, for her friendship, boundless enthusiasm, critical acumen, and excellent collaboration. To Anthony McCosker, for his friendship through the ups-and-downs of the past few years, his intellectual generosity, astute critical eye, and for acting as a constant sounding board over many good coffees at Swinburne as I worked through various ideas early on—ありがとうございます. And, to Jenny Kennedy, for her friendship, collaboration, and encouragement (especially in the latter stages of writing and editing).

My biggest debt of gratitude goes to my four most favorite people: Karen, Laz, Max, and Sunday—thank you for everything! This book would not have been possible without you. Your endless love and support are what got me through.

Bits and pieces of this book have appeared elsewhere, and have been reproduced here with permission.

Parts of the Introduction were adapted from the following: Rowan Wilken, "Locative Media: From Specialized Preoccupation to Mainstream Fascination," *Convergence: The International Journal of Research into New Media Technologies* 18(3) (2012): 243–247; Rowan Wilken and Gerard Goggin, "Locative Media—Definitions, Histories, Theories," in Rowan Wilken and Gerard Goggin (eds.), *Locative Media* (New York: Routledge, 2015), 1–19.

Chapter 3 has been adapted from the following: Rowan Wilken, "'Places Nearby': Facebook as a Location-Based Social Media Platform," *New Media & Society* 16(7) (2014): 1087–1103; Rowan Wilken and Peter Bayliss, "Locating Foursquare: The Political Economics of Mobile Social Software," in Rowan Wilken and Gerard Goggin (eds.), *Locative Media* (New York: Routledge, 2015), 177–192; Carlos Barreneche and Rowan Wilken, "Platform Specificity and the Politics of Location Data Extraction," *European Journal of Cultural Studies* 18(4–5), 497–513; and Rowan Wilken, "The De-gamification of Foursquare?" in Michele Willson and Tama Leaver (eds.), *Social, Casual and Mobile Games: The Changing Gaming Landscape* (London: Bloomsbury, 2016), 179–192.

One section of Chapter 4 drew from Rowan Wilken, "A Community of Strangers? Mobile Media, Art, Tactility and Urban Encounters with the Other," *Mobilities* 5(4) (2010): 449–468; and from Rowan Wilken, "Proximity and Alienation: Narratives of City, Self, and Other in the Locative Games of Blast Theory," in Jason Farman (ed.), *The Mobile Story: Narrative Practices with Locative Technologies* (New York: Routledge, 2013), 175–191.

And, a small part of Chapter 5 was drawn from the following article: Rowan Wilken, "Mobile Media and Ecologies of Location," *Communication Research and Practice* 1(1) (2015): 42–57.

Many of the ideas in this book were tested out on different audiences at various symposia and conferences, including the Association of Internet Researchers (AoIR), the Australian and New Zealand Communication

Association (ANZCA), Crossroads in Cultural Studies, Geomedia, and the International Communication Association (ICA). I thank the organizers of these events for these opportunities, and the constructive feedback offered by those present. The ideas in this book have also benefited significantly from the input of those who participated in the "Social Lives of Locative Media" symposium I was fortunate enough to be able to host at Swinburne University of Technology in July 2014.

This book is an output from a program of research under an Australian Research Council (ARC) Early Career Researcher Award (DECRA)—DE120102114, "The Cultural Economy of Locative Media." I gratefully acknowledge the ARC's generous financial support.

Rowan Wilken, RMIT University

37°48'26" S 144°57'53" E

37°49'19" S 145°02'15" E

Cultural Economies of Locative Media

Introduction

In Sydney Pollack's first film, *The Slender Thread* (1965), college student Alan Newell (Sidney Poitier) is volunteering during the night shift at a Seattle crisis call center when he receives a telephone call from a suicidal Inga Dyson (Anne Bancroft), who has swallowed a handful of barbiturate pills. What follows is an increasingly desperate bid to try to locate her in order to save her life; it is a race against time involving Newell, his psychiatrist supervisor (Telly Savalas) and support staff, the police department, and various telephone exchange operators and engineers. The phone conversation becomes a cat and mouse game between Newell and Dyson: Newell is eager to glean any piece of information that might help to locate his caller; Dyson is careful not to disclose her whereabouts. At one point, while telling a part of her story to Newell, Dyson stops abruptly and declares, "I shouldn't have told you about the office, you might try and trace me." This is precisely what Newell sets about doing, surreptitiously placing a call on the crisis clinic's second phone line to the local switchboard operator requesting "an emergency trace." Following some detective work on the part of a telephone exchange engineer, Newell receives word from the switchboard supervisor that "[i]n tracing your call, we find it is coming through an outlying switching station. We're sending a man out to complete the trace." In the 1960s, telephone calls were routed and connected through a series of electro-mechanical switches; so long as the caller did not hang up the phone, the call could be traced back by hand through these switches in order to determine the exchange the call was coming through and, from there, the actual line, which would then give the caller's actual location (Dobyns 2015). Newell's increasingly desperate attempts to keep Dyson on the line to facilitate the trace, and his efforts to ascertain her location ("What's that [sound]? Was that a plane?"), are punctuated by Dyson's retelling of further bits and pieces of her story, and the frantic efforts of telephone engineering staff to reach the outlying exchange building. One worker can't be contacted, so another is pulled from a date with his girlfriend at a local pool hall to complete the job. This second worker borrows a motorbike, which runs out of fuel en route; he eventually makes it to the exchange building and completes the trace in the nick of time, successfully locating Dyson just as she drops the phone receiver and begins to drift in and out of consciousness.

On one level, *The Slender Thread* reads four decades later as something of a cinematic curiosity, a historical timepiece or snapshot, not so much of a simpler age but, to a contemporary sensibility, of a period in recent technological and human history that is as complicated as ours yet altogether stranger because so seemingly unfamiliar. Would all those people—crisis counselors, telephone exchange staff, multiple emergency services workers—really have become enrolled in a coordinated search to pinpoint a single caller's location? And, did it really take that long—several tension-filled hours—to determine the precise location from which one phone call originated? The dramatic emphasis that is given to this search seems unlikely—absurd, even—to us now, given that, with modern telephone exchange systems, mobile handsets, and associated infrastructures (cell tower triangulation, GPS trilateration, etc.), tracing a caller's location is an almost trivial exercise—one that, in most cases, can be completed instantaneously or within minutes of a call being placed.

On another level, though, *The Slender Thread* continues to resonate after all these years in a number of important ways. Fundamentally, it speaks to us of the enduring importance of location to all forms of telecommunication, whether these be fixed-line or mobile. In addition, when thinking about the importance of location to various forms of communication, it speaks to us of the need—one echoed in calls by a number of scholars (Bowker et al. 2010; Horst 2013; Parks and Starosielski 2015a, 2015b; Star and Ruhleder 1996)—to pay due and careful attention to the infrastructures that enable and facilitate communication technologies and their uses. And, it highlights the importance of historicizing our contemporary accounts of technology emergence and use (Gitelman 2006), and of being attuned to the longer histories of the "compelling tangle of modernity and technology" (Misa 2003). It also reminds us of the important fact that, when it comes to determining where those on the other end of the phone are, pure geographical location is rarely the primary consideration; rather, the importance and complexity of location "comes from the fact that location intertwines with other relevant aspects of context" (Arminen

2006, 322), including, to name just two, the identity of the persons who are party to the interaction, and the information that is associated with, and captured and conveyed by, these interactions.

The aim of this book is to explore and make critical sense of the enduring importance of location and, more specifically, of the importance of location to mobile devices. The book provides a detailed critical account of contemporary location-sensitive mobile media and their significance, exploring the ideas, technologies, practices, contexts, corporate arrangements, and power relations that define and shape them.

In the US context, steady growth since 2010 in the take-up of smartphones, and with this the wider adoption and acceptance of locative media services, is clearly evident in research undertaken by the Pew Research Center. Back in 2010, in their Internet and American Life study, Pew researchers found that "only four percent of adults online use location-based services," which they define as "sites and mobile apps like Foursquare and Gowalla, programs that allow users to 'check-in' to various locations and tell their friends where they are" (Horn 2010). The following year, 55% of respondents with smartphones reported that they used phones "to get real-time location-based information" (which is clearly asking about wider modes of registering location than via check-in services alone) (Zickuhr 2012). This figure rose to 74% by 2012 (Zickuhr 2012), and 90% by 2016 (Pew Research Center 2016). In summarizing their earlier (2012) findings, Kathryn Zickuhr (2013) observes that Pew researchers observed a "modest drop in the number of smartphone owners who use 'check in' location services," a downward trajectory that has since continued. Nonetheless, it is also clear from the results that "the role of location in digital life is changing as growing numbers of internet users are adding a new layer of location information to their posts, and a majority of smartphone owners use their phones' location-based services" (Zickuhr 2013). In overall terms, Zickuhr (2013, n.p.) concludes, "these trends show the ascent of location awareness."

The upward trends in increased location awareness that the Pew Research Center identifies are by no means restricted just to the United States. Location services are heavily reliant on mobile broadband, and, as the International Telecommunications Union's annual ICT Facts and Figures reports reveal, global mobile broadband subscription rates are growing exponentially. These have "climbed from 268 million in 2007 to 2.1 billion in 2013" (ITU 2013), which converts to an average annual growth rate of 40%, with subscription rates climbing again to 3.6 billion by the end of 2016 (ITU 2016) and still further to 4.3 billion globally by the end of 2017 (ITU 2017).

Such research is instructive in highlighting the growing global significance of what is the core concern of this book: location-sensitive mobile media—hereafter referred to as *locative media*.

WHAT ARE *LOCATIVE MEDIA*?

Some orientation is warranted here, given the complications associated with both the component terms in the phrase *locative media*. I begin, therefore, by reflecting on these two terms, both individually and in combination.

When it comes to the term *media*, in this book I follow Lee Humphreys's lead in conceiving of media as "those tools and channels that connect people across time and space and allow for the sharing of meaning" (Humphreys 2018, 8). In developing this account of locative media, I am also sensitive to Roger Silverstone's (2005) argument that media is enrolled in larger processes of mediation that require us to be attentive to the social and "institutionally and technologically driven and embedded" forms of communication:

> Mediation, as a result, requires us to understand how processes of communication change the social and cultural environments that support them as well as the relationships that participants, both individual and institutional, have to that environment and to each other. At the same time it requires a consideration of the social as in turn a mediator: institutions and technologies as well as the meanings that are delivered by them are mediated in the social processes of reception and consumption. (Silverstone 2005, 189)

Yet, any conception of media, in this context, is complicated. As Gerard Goggin and I have noted elsewhere, "the media part of this [locative media] couplet is expanding almost beyond comprehension," such that locative media involve "global positioning satellites (GPS), cellular mobile phones, location-based services (LBS), [and] mobile social software, social networking applications, and so-called check-in applications" (Wilken and Goggin 2015, 2). Locative media, as is argued later in this book, are also shaping up as the "harbinger of the emergent media of our time, from big data to drones, from the Internet of Things to logistics, all with their urgent cultural, social, and political implications" (Wilken and Goggin 2015, 2).

Within earlier studies of mobile telephony, geographical considerations (of place and of location) were often viewed as subsidiary to other communicative considerations (for discussion, see Wilken and Goggin 2012a, 2015). This has since changed significantly. As Pew researchers observe in their 2016 report, people increasingly use their smartphones for more than voice calls and texting (Pew Research Center 2016). One argument of this book, to take the Pew findings a step further, is to suggest that location-sensitivity is a constitutive feature of all smartphones and now fundamentally underpins much contemporary smartphone use.

If *media* in the *locative media* couplet is tricky due to its expansion in meaning, *location* is equally complicated. The concept of location has developed according to a diverse and complicated set of etymological trajectories that include legal use (with location understood as the action of letting for hire) and grammatical use (where it refers to a particular case form) (Simpson and Weiner 1989, 1081–1082). It also refers to land settlement practices, as well as to processes of emplacement and "the action of discovering, or the ability to discover or determine, the position of a person or thing" (Simpson and Weiner 1989, 1082). It is these last two senses that constitute the general understanding of the term and that most inform how the term is employed in relation to mobile media technologies.

According to Fred Lukermann (1961), location is both a foundational concept for geography *and* one that carries important general significance. From the Ancient Greek poets onward, he writes, "how to describe 'where something is' becomes idiomatic in Western culture" (Lukermann 1961, 197). And yet, location, at least as it is generally understood, has since come to be understood as a subsidiary concern to the more encompassing concept of place. This is a perspective that is captured in Ed Relph's (1986, 3) description of place as "location plus everything that occupies that location seen as an integrated and meaningful phenomenon." More recently, however—and commensurate with the rise of location-enabled mobile communication technologies—location has come to be recognized as having taken on increased conceptual importance in its own right. Adriana de Souza e Silva and Jordan Frith encapsulate this renewal of interest in the concept of location as follows:

> The popularization of location-aware mobile technologies not only highlights the importance of location, but also forces us to re-think how location has been traditionally conceptualized. Locations are still defined by fixed geographical coordinates, but they now acquire dynamic meaning as a consequence of the constantly changing location-based information that is attached to them. (de Souza e Silva and Frith 2012, 9)

Thus, they argue, where locations were once seen as "places deprived of meaning" (or perhaps, whose meaning was dependent on other concepts and phenomena) they can now be seen as taking on "complex, multifaceted identities that expand and shift according to the information ascribed to them" (10).

Locative media is useful in this context for the simple reason that it is economical and expansive, but also precise. While multiple alternative terms have been coined to capture many of these same developments—such as "geomedia" (Fast et al. 2018; Lapenta 2011; McQuire 2016), "spatial media" (Kitchin, Lauriault, and Wilson 2017), and "geographies of communication" (Adams 2009; Falkheimer and Jansson 2006)—*locative media* is the term given preference in this book. These two words capture a great deal while retaining a sense of the term's very particular history, which is anchored in the field of new media arts (Tuters 2012; Zeffiro 2012), and which has long been at the vanguard of exploring the experimental and creative possibilities and the critical implications of locative media technologies (see Chapter 5 for detailed discussion). *Locative media* is a term broadly defined as "media of communication functionally bound to a location" (Wikipedia 2017a), with these media exploring information, data, sounds, and images about a location. The efficacy of this term for describing exploratory developments in mobile media technologies is a product of the meanings carried by it (Tarkka 2010). Media arts theorist Karlis Kalnins, the person generally credited with coining the term "locative media," is understood to have been drawn to the word "locative" based on his knowledge of "languages such as Latvian and Finnish with their several locative cases corresponding roughly to the preposition 'in,' 'at,' or 'by,' and indicating a final location of action or a time of the action" (Tarkka 2010, 134). This etymological preference on Kalnins's part is more than mere semantics; it is a deliberate move that, for him, strategically repositions media arts practice by shifting the emphasis off the site of action (actual places or locations) and onto the agency and actions of subjects in the temporal dimensions of these actions. In using "locative media" here, I acknowledge the significance of Kalnins's linguistic preference for "locative" over "location." However, the position I take in this book is that the two—actions *and* the sites in which these temporally based actions occur—are best kept in productive tension, so as to account for the various technological and other infrastructures that mediate our locationally situated technosocial interactions.

Initially considered a somewhat specialized pursuit or preoccupation (especially within new media arts, as noted), locative media are now very much shaped by mainstream uptake and have become the focus of increased consumer fascination (Wilken 2012b). This includes wide user familiarity in using an array of mobile, location-related features and functions: satellite navigation systems, mobile maps, location-based games, social networking services, and various location apps on iPhones and smartphones that use and rely on the registering of geolocation. Such use has become increasingly commonplace. Every day, tens of millions of mobile users identify and register their location in various ways via the functions embedded within or downloaded onto their mobile devices. For example, the images and videos we tag, upload, and share to sites such as Snapchat, Instagram, WhatsApp, Flickr, Facebook, and Kakao, are now typically geocoded. Positional information is also a core component of dating and hook-up apps, such as Grindr and Tinder. And, as the vignette that opened this chapter demonstrates, accurate, real-time geolocation data are a vital resource for modern emergency services agencies.

With this democratization of locative media, outside of the field of media arts there has been a wide flowering of critical interest in, and a growing body of interdisciplinary scholarship on and around, locative media. Furthermore, within this broader literature, myriad considerations are taken up that range across (to name only a few) analysis of how locative technologies mediate the relationship between technology use and physical/digital spaces; questions of power (and its uneven distribution) (de Souza e Silva and Frith 2012, 136–161); the capacity for social inventiveness (or otherwise); personal identity formation, or what—*pace* Erving Goffman—has been termed "the presentation of location" (de Souza e Silva and Frith 2012, 162–184) and the "spatial self" (Schwartz and Halegoua 2015); the monetization efforts of locative media firms (Frith 2015b); the importance of mobile-human interfaces and the need for more thorough theorizations of these relationships (de Souza e Silva and Frith 2012, 23–108; Farman 2012); privacy (de Souza e Silva and Frith 2010, 2012, 111–135; Gordon and de Souza e Silva 2011, 133–154); and, (post-)phenomenological reflections on the use of locative mobile media (L. Evans 2015; L. Evans and Saker 2017; Richardson and Wilken 2009).

KEY DEVELOPMENTS THAT FUELED THE GROWTH OF LOCATIVE MEDIA

Three major technological developments toward the end of the first decade of the twenty-first century drove the wider uptake of, and flourishing critical interest in, locative media. These developments put locative media at the center of contemporary cultural and social dynamics, fueling interest in location-associated data ("datafication"), and contributing to a profound reshaping of our understanding of everyday engagements with location, communication, and social interaction.

The first of these involved the democratization of digital mapping technologies. In the wake of Google's embrace of geolocation services in 2005, mainstream interest in and uptake of locative media services, especially maps, flourished. The result has been such that "consumers are now well accustomed to using sat nav devices in their cars or while walking, Google Maps on desktop and laptop computers and mobile devices, and geoweb, geotagging, and other mapping applications" (Wilken and Goggin 2015, 5; see also A. Crawford and Goggin 2009; Thielmann 2010, 5).

The second set of developments followed from the launch of the Apple iPhone and Google's Android operating system in 2007–2008, which led to extraordinary growth in smartphone take up and use and the consolidation of the mobile internet (see Hjorth, Burgess, and Richardson 2012; de Souza e Silva and Frith 2012, 96–97). The rise of the smartphone— the "iPhone moment," as Gerard Goggin (2011, 181) refers to it—led to a number of additional and significant developments: an acceleration of "the trend toward the crossover between Wi-Fi (wireless internet) and cellular mobile networks and devices," such that handsets, applications, and users now switch with ease among networks (Wilken and Goggin 2015, 6); an associated international "evolution of network architectures and infrastructures, as telecommunications and mobile networks—especially next generation networks and 4G and 5G mobile networks—merged with Internet protocol and data networks" (6); and, the rise of a vibrant "app economy" servicing platform owners, mobile application developers, and smartphone end-users (Goldsmith 2014; Wilken 2018a). As OpenSignal's Sina Khanifar has observed, "the amount of information you can get from location [via smartphones], the amount of things that you can correlate it to" is what made the "iPhone moment" so significant, especially for app developers (Sina Khanifar, pers. comm., 2013).

The third set of developments concerned the "enormous growth in personal, private, and machine-based information and processing [that have] been associated with a wide range of consumer and enterprise technologies and networks, which add significantly to mass personalized user, device, and network data concerning location" (Wilken and Goggin 2015, 6). These developments are often associated with: RFID (radio-frequency identification) technologies (Frith 2015a, 2019; Hayles 2013; Rosol 2010); the networked interconnection of computing devices that are embedded in everyday objects known as "the internet of things" (Bunz and Meikle 2018; Greengard 2015); sensors and microprocessing chips that capture and transmit data, "such as sound waves, temperature, movement, and other variables" (Swan 2012, 217); and globe-spanning industrial and software-related logistics industries (Rossiter 2016, 2015). Increasingly miniaturized computing chips and sensor technologies, coupled with advances in machine learning capabilities (Mackenzie 2017), have thus come to play vital roles in *connecting* people, goods, and other material things, and in facilitating and tracking the *movement* of people (both indoors and out) and of goods and other material things.

These three sets of developments have had a profound impact in fostering the democratization of, and opening up of access to, geolocation services and associated infrastructures. Not only have location and location-awareness become increasingly central to our contemporary engagements with the internet (the geoweb) and mobile media, there has been the suggestion that "unlocated information will cease to be the norm," and location will become a "near universal search string for the world's data" (Gordon and de Souza e Silva 2011, 19–20). As Malcolm McCullough (2006) has put it, information "is now coming to you [. . .] wherever you are" and "is increasingly *about* where you are."

The importance and impact of this information is now being felt across a range of spheres of modern, industrialized life. To cite just a few examples, location data have been and continue to be central to the following: the growth of search and social media, and ongoing transformations within the media and advertising industries (with their growing reliance upon automated recommender systems and geodemographic profiling); global logistics systems; rapid growing interest and investment in "smart cities"; cloud computing; precision agriculture; autonomous vehicle development; emergency relief efforts; spatial big data analytics; and, of course, corporate-government surveillance and the military-industrial complex more broadly.

"Where Are You?" and Whenabouts in the Age of Mobile Phones

This explosion in interest and wider use of smartphone-enabled locative media in particular should not, however, be taken to suggest that interest in the question of where phone users are located is necessarily a new one. On the contrary, researchers of mobile communication have long been interested in the "Where are you?" question (for discussion, see Wilken and Goggin 2012a, 16–17).

Moreover, the preceding account of key developments aiding determination of location should not be taken to suggest that the ability to *gather* location data is necessarily new. While consumer, corporate, and government embrace of smartphone-enabled location-based services is comparatively recent, the use of cell phones as positioning technologies is well established. For example, Gerard Goggin (2006, 195) notes the three broad ways "of locating a handset or other user equipment with cellular networks" in the pre-smartphone era of mobile communications. The first took advantage of cellular radio design, and permitted the user's handset to be identified as being located within a particular cell (196; for detailed discussion, see Ling and Donner 2009, 31–33; Soltani and Gellman 2013a); the second involved measurement of the time signals it took to travel from a handset to two or more network base station transmitters (known as triangulation) (196); and, the third, GPS, permitted the "calculation of position based on propagation delays of different transmissions" (known as trilateration) (196).

What has changed with the arrival of the smartphone is that the determination of end-user location can be calculated with even greater accuracy. In addition to the methods described in the preceding, the position of smartphones can be calculated through their interactions with 3G, 4G and, soon, 5G networks (Goggin 2006, 195–197), and via Wi-Fi networks (de Souza e Silva 2013, 117), which, in the words of one writer, "have become the navigational beacons of the 21st century" (S. Gallagher 2014). Thus, while we as users play a vital role in disclosing location information through use—"much of the digital content we produce and share is or may easily be geocoded" (Leszczynski 2017, 235)—our devices are also conveying this information, often in ways that are unbeknownst to us (Leszczynski 2015, 970). For instance, Steven Vaughan-Nichols (2011, n.p.) explains,

> Android Location Services periodically checks on your location using GPS, Cell-ID, and Wi-Fi to locate your device. When it does this, your Android phone will send back publicly broadcast Wi-Fi access points' Service set identifier (SSID) and Media Access Control (MAC) data.

This, Vaughan-Nichols (2011) points out, is not just something that Google does; rather, it is standard "industry practice for location database vendors." What is more, smartphones are also packed full of sensors—including accelerometer, gyroscope, magnetometer, proximity, barometer, and air humidity sensors, GPS, and Wi-Fi and Bluetooth locators, among others—several of which play a role in determining a phone's location. In this way, smartphones have become "drone devices" that occupy a "continuous background presence" where they operate as minitiarized, distributed, always on, "monitoring 'assemblages'" (Andrejevic 2015, 197; Andrejevic and Burdon 2015).

Thus, location-sensitive smartphones are considered to be categorically different from previous mobile media in that they more explicitly "draw information from the physical surroundings" (de Souza e Silva and Frith 2010, 507). For Adriana de Souza e Silva and Jordan Frith (2010), this articulation of location and data, which is characteristic of locative media, carries three implications:

(1) it allows users to "interact with previously existing local information" (507);
(2) it allows users to "create local information that might be shared with others in the vicinity," whether through reviews of sites or service or through other forms of "annotations" (507; see also, Thielmann 2010, 2; Gordon and de Souza e Silva 2011, 40–58); and,
(3) it allows "users to select information from the surrounding space they want to interact with" (de Souza e Silva and Frith 2010, 507).

This, in turn, has led to end-user location and location-based services becoming "of intense interest to the cell phone, wireless, and mobiles industries" (Goggin 2006, 196), as well as the advertisers and marketers, and the major players of the technology industry, such as Google and Facebook. The reason is simple: smartphones "gather unprecedented amounts of longitudinal data on their users' locations," all of which facilitate "tailored retail and consumer services, lifestyle profiling and mapping, and surveillance" (Wilken and Goggin 2015, 6). These forms of location data gathering raise manifold questions that this book seeks to examine.

GENERATIONS OF LOCATION-SENSITIVE MOBILE SOCIAL NETWORKING

Across the various chapters of this book, considerable attention is paid to mobile locative social networking services over different time periods. Given this emphasis, it is valuable to say a few words here about the evolutions that these services have undergone since the first decade of the 2000s. In addition to specific waves of technological developments—including the advent of the smartphone with incorporated mapping capabilities—that gave rise to locative media, there have also been specific stages that have marked the rise of *location-sensitive mobile social networking* (LMSN) services. Characteristic of these services is the ability to register's one's physical presence at a particular location or venue, which is then communicated to one or more people who are either within an individual's social network or are fellow users of the same service who are unknown to an individual.

Since around the year 2000, contemporary smartphone-enabled location-based services can be described as having passed through at least three iterations or generations. "First-generation" location-based services required the active registering of one's location by end-users, often in the form of "check-ins." Several of these pre-date the release of the first iPhone and were thus designed to operate on older mobile handsets and associated infrastructures. A pioneering early example was Dodgeball (cofounded by Dennis Crowley in 2000, and subsequently sold to Google in 2005) (see Humphreys 2010, 2007). Other, related services followed, such as Loopt (founded in 2005), Whrrl, Brightkite, and Gowalla (all founded in 2007), and Foursquare (cofounded by Dennis Crowley in 2009, the same year that Google discontinued Dodgeball), and Chinese Foursquare clone, Jiepang (founded in 2010). Of these, only Foursquare continues to survive as an independent operation: Brightkite sold to HDmessaging (formerly Limbo) in 2009, Whrrl to Groupon in 2011, Gowalla to Facebook in the same year, and Loopt to Green Dot in early 2012; and, having undergone a significant redesign in 2013, Jiepang closed in 2016. The locative media landscape changes quickly.

"Second-generation" LMSNs, also known as "ambient social location" or "social search" applications (Lee 2013, 27–28), involved "passive" location disclosure, tracking, and compatibility pairing of end-users. Indicative of these "second-generation" LMSNs were applications such as Glancee and Sonar (both founded in 2010), and Highlight and Banjo (both founded in 2011). The first of these, Glancee, tracked a user's location in the background, and linked to the user's Facebook and Twitter accounts, to show the user "people who are using the app [nearby] and their shared social graph interests and Facebook picture" (P. Burns 2012); it also included a "radar" function to reveal their physical proximity (Lee 2013, 27). As with those of the first generation, very few second-generation applications have survived as ongoing independent operations—not least due to consumer perception that they were inherently "creepy." Glancee was purchased by Facebook in 2012 and closed; support for Sonar app was halted in 2013 (Ha 2013); and, Highlight ceased operation in 2015 following a "talent acquisition" by Pinterest (Lynley 2016).

Third-generation" LMSNs, arguably the dominant form at time of writing, are more broadly defined location-based services that involve what I am here referring to as *ubiquitous geodata capture*. For these services, location remains fundamental to their operation, but is integrated at both the front end (the interface) and the back end (algorithmic processing, database population, monetization efforts, and so on).

The Uber app provides a useful example of a "native" third-generation service in that geolocation and geocoded data are seamlessly integrated into, and vital to, the operation of the service at all levels. Of the specifically location-focused applications discussed earlier, the two that have survived—Banjo and Foursquare—have done so by undergoing significant transformation from first- or second-generation LMSN applications to third-generation location-based services (LBS), or platforms, where location has become a core component of a wider suite of offerings that are specifically geared around real-time data analytics. And, in addition to newer entrants like Uber, and stalwart LMSN services like Banjo and Foursquare, established search and social media firms, such as Facebook and Google, have also effectively become "third-generation" location-based services platforms, insofar as they have both reshaped, to varying degrees, their operations (and, in Facebook's case, those of its subsidiaries) so as to facilitate the integration, at a range of scales, of ubiquitous geodata capture across their operations (see Barreneche 2012a; Barreneche and Wilken 2015; Wilken 2014a).

What these developments reveal is a significant shift over time in the way that location has been registered, from the active check-in of first-generation services, to passive logging of second-generation services, to ubiquitous geodata capture of third-generation services. While all three phases are integral to the story this book tells about locative media, a key larger argument of this book is that it is the thorough integration of location data capture capacity that we see with contemporary services that highlights the continued value and significance of location. While often no longer clearly visible within many social media and other related services, location, locatability, and location data now form vital ingredients in many new technological developments, and will likely long remain so as these technologies evolve and others emerge.

FRAMING CULTURAL ECONOMIES

The approach that gives structure and shape to the analysis of locative media in this book is that of *cultural economy*, broadly conceived. This is an approach with a rich, if at times fraught, history (see Amin and Thrift 2004, xv–xviii; Flew 2009; C. Gibson and Kong 2005). As Chris Gibson and Lily Kong (2005) point out, cultural economy is a decidedly polyvalent term; it is "umbrella-like, used to embrace a range of different understandings" (Anheier and Isar 2008, 3) that depend on disciplinary and critical orientation.

As it is applied within the pages of this book, cultural economy is taken less as a formal method or systematic approach and more as a general orientation—a point of entry for exploring "the ways in which the 'making up' or 'construction' of economic realities is undertaken and achieved; how these activities, objects and persons we categorize as 'economic' are built up or assembled from a number of parts," and how they interact (du Gay and Pryke 2002, 5; see also Goggin 2014, 2006; Gopinath 2013). It provides a broad yet productive frame through which to view locative media in all its industrial, end-user, and regulatory complexity. In other words, what this frame permits is a multi-perspectival account of locative media and, among other things, their production and consumption, and the regulatory and policy forces that interact with and shape them.

In taking a cultural economy approach, however, I do not wish to draw or maintain strict conceptual or disciplinary boundaries between cultural and political economy. While cultural and political economy approaches have at times been opposed to each other—sometimes antagonistically so (Garnham 1995; Grossberg 1995)—there are also significant consistencies between them, with a number of points of intersection and overlap (Best and Paterson 2010; Fenton 2007). Both, for instance, claim an interest in the formation and maintenance of power and in tracing uneven power differentials (Amin and Thrift 2004, xxi; Golding and Murdock 2000, 73). My adoption of a more catholic approach to cultural and political economy is in part a product of my own education and research trajectory, which has seen me shift across media and communication and cultural studies and back again. An inclusive approach also accords with a growing recognition that there is more to be gained from embracing composite critical tools, especially when dealing with fast-moving research targets like mobile and locative media. To adopt Lily Kong's words (cited in C. Gibson and Kong 2005, 556), "the question is not whether to rediscover economics or to go with the cultural, it is how to do both at the same time in ways that recognize the political significance of these intersections and provide a critical purchase" on locative media.

Thus, the decision to employ *cultural economies* as the framing concept for this book is a result of a number of epistemological considerations, raised in the preceding discussion, as well as a number of practical considerations. While I have a great deal of sympathy for Jacqueline Best and Matthew Paterson's (2010) efforts to bridge cultural and political economic approaches, *cultural economies* has a certain concision to it that is not captured in hybrid, composite terms, such as their "cultural political economy."

The concept of cultural economies also forms a productive rubric insofar as it more explicitly references considerations of end-use of locative media that are taken up in Part II and in Chapter 8 in Part III of this book. Ash Amin and Nigel Thrift (2004, xviii) suggest that cultural economy is, in certain respects, "more directed towards actual practice." They argue that the "intellectual bounty" of this focus on practice, both within industry and among end-users, has been a "better appreciation of the economy as cultural practice and of culture as economic practice," such that, for instance, we have gained a fuller understanding of "learning and knowledge acquisition in firms, and the emotional investments made by consumers in mundane but crucial practices such as shopping or investing" (Amin and Thrift 2004, xviii). An appreciation of practice can also assist in developing an understanding of the talent and knowledge acquisition and labor processes of location-related tech firms (see, for example, Marwick 2013; Neff 2012) and of the emotional investments made by consumers who engage with the products of these firms, such as LMSN apps like Foursquare (see, for example, Frith 2013; Halegoua, Leavitt, and Gray 2016).

Notwithstanding the broad cultural economies focus, at other points in the book (such as Part I), political economy approaches to considering questions of power and control, ownership, and revenue models are given greater emphasis. The political economy of communication is an established and well-tested approach (Hardy 2014; Mosco 2009; Wasko 2004) that has been applied not only to the analysis of regulated broadcast media industries, but also to the study of search (Mosco 2014; Van Couvering 2011), mobile (Goldsmith 2014), locative (Wilken and Bayliss 2015), and social media industries (Albarran 2013; Wilken 2018a), as well as to new forms of media consumption and distribution (Lobato and Thomas 2015). As this list reveals, there has been considerable effort to explore the political economy of "new media"—or what Dwayne Winseck (2011, 7) calls "the multiple economies of network media." One productive, early account of what political economy can bring to the study of new or networked media was developed by Robin Mansell (2004, 96) when she called for "a revitalization of research on new media in the tradition of political economy." Drawing on the work of earlier political economy theorists and critics, Mansell argues that "any political economy of new media must be concerned with symbolic form, meaning and action as it is with structures of power and institutions" (98) and with the ways that "power is structured and differentiated, where it comes from and how it is renewed" (99).

Examining the revenue models of location-based services, and how they go about achieving a sustainable economic basis, is a key issue that a political economic analysis can address. Mansell highlights the ways in which scarcity remains an important dynamic, despite "the abundance and variety of new media products and services" (97) often emphasized by the technology trade press. More precisely, she writes, "[t]he production and consumption of new media in their commodity form means that scarcity has to be created by, for example, [the] creation and sale of audiences" (98). It is this production of artificial scarcity that underlies corporate strategy for firms such as Foursquare, Facebook, and others in developing a sustainable business model based on their proprietary database of locations and, importantly, on users' patterns of interactions with those locations, particularly via data accrued through third-party access to its application programming interface (API). Political economy approaches are also useful insofar as they afford a way of understanding LBS firms' relationships with and significant reliance on their user bases and the "concrete consequences" of patterns of power and ownership "for the work of making media goods" (Golding and Murdock 2000, 84).

Thus, it is worth reiterating that, while cultural economy and political economy are given different emphases at different times, with each coming into and out of focus at different points, these are not viewed here as competing or mutually exclusive frameworks; rather, in this book they form a unified approach that, for simplicity's sake, will be collectively referred to as the "cultural economies of locative media." Such a unified approach is important and necessary if we are to make sense of locative media in all its industrial, end-user, and regulatory complexity.

QUALIFICATIONS

Having described the theoretical approach that guides the analysis of this book, it is prudent at this point to also note a number of key discursive "silences" (Macherey 2006) within, and limitations of, the present text. These are important to acknowledge, especially in light of Lisa Gitelman's (2006, 1) observation that any work of media history is inevitably always partial by virtue of the fact that it takes on particular foci depending on the methods and approaches employed, and material not covered. The explicit focus of this book is (a) on mobile location-related technologies that have been developed, in the main, by US-based companies, and (b) Anglo-American, European, and Australian cultural contexts, often involving predominantly white early adopters in big cities. While it has not been intended, this double focus imposes a number of restrictions on the present study. Admittedly, these restrictions have proven helpful insofar as they have limited and focused the scope of the present study. But these same limits also mean that what is portrayed here is only one particular set of reflections on mobile locative media, and has the potential to exclude or overlook others and alternative approaches to this topic. It does not, for instance, heed calls to "internationalize" and "de-Westernize" communication studies and to move beyond Anglophone paradigms (see, for instance, Goggin and McLelland 2009; Thussu 2009; Wang 2013; Willems and Mano 2017). While it has been beyond the scope of this book to engage with this issue in detail, it is one that I have sought to respond to elsewhere (see Wilken, Goggin, and Horst 2019). In addition, the prevailing focus within much existing work on mobile location-sensitive social media on predominantly white early adopters who are based in big cities risks excluding the experiences of many other communities of users from these accounts (e.g., Brock 2012; Frith 2017), and risks eliding the significant power dynamics that shape use of these technologies (e.g., Noble 2018; Pasquale 2015). Again, while it has been beyond the scope of this book to engage with these issues in detail, it should be noted that important work is being done (within media anthropology, communication studies, internet studies, and elsewhere) in responding to these issues. There nevertheless remains a pressing need for further detailed research on the uses (Albury et al. 2017), pitfalls (e.g., Akil 2016) and promise (e.g., Brock 2009) of mobile location and social media technologies for specific minority groups and for wider communities of users beyond those addressed in this book.

BOOK STRUCTURE AND CONTENTS

And so, finally, to an explanation of the structure of the book: This book is composed of three interconnected but stand-alone parts. The intention in structuring the book in this way is that, collectively, these parts build a composite (rather than progressive) argument about locative media. The three parts can be understood as functioning like rooms within an art exhibition. All the rooms in the exhibit are connected. However, each room is set up to frame the objects and ideas it contains in a particular way, and each asks for its own set of reflections (Griffiths 2016, 217). It is only by taking in all the rooms within the exhibit—in whichever order the reader-visitor desires—that one arrives at a fuller and deeper understanding of the book-exhibition as a whole and the themes and issues and arguments that it explores.

Included at the beginning of each part within the book are introductions that explain the broad aims and methods employed in each part, and which contain more detailed chapter summaries than those that follow here.

The aim of this Introduction has been to map the overall terrain that the book traverses, and to orient the reader within this terrain.

Part I, "The Topography of Locative Media," includes three chapters that combine to provide a detailed picture of the business side of locative media. Chapter 1 examines the larger mobile location ecosystem and its industrial formation; through this examination, a clearer picture of the industrial composition of location-based services (LBS) emerges. Chapter 2 considers the importance of maps and mapping assets within the larger location-based services ecosystem. Chapter 3 examines the geolocation data capture efforts and business and revenue models of two commercial businesses now central to the contemporary settlement of locative media: Foursquare and Facebook.

The focus of Part II is on artistic and everyday cultural practices involving location-sensitive mobile technologies. Chapter 4 looks in detail at three locative media arts projects—by Blast Theory, Josh Begley, and Julian Oliver—and how each, in its own way, utilizes location technologies for thinking through aesthetics and its relationship to politics. The next two chapters give detailed consideration to quotidian uses of location technologies. Chapter 5 looks at how any one locative media application is routinely used alongside and as intermingled with a wide array of other applications that also include various forms of location functionality. Chapter 6 examines how urban spaces and places are explored, catalogued, and communicated through location-sensitive smartphone applications, and how these communicative practices are entwined with processes of individual identity construction and performance.

The final part of the book, Part III, contains two chapters that take up and explore a complicated set of issues around how location data are extracted, by whom, and to what ends, and the many implications this has for individual consumers, as well as policy debates concerning data storage and privacy. Chapter 7 explores these issues via two case studies: one on Google's Street View Wi-Fi data-gathering program, and the other on the US National Security Agency (NSA) and its far-reaching surveillance program, as revealed through the Snowden papers. Chapter 8 examines one of the most pressing yet thorniest of policy issues pertaining to social media and mobile communications use for companies, end-users, and regulators alike: privacy.

The book concludes with a short Conclusion that revisits the terrain the book has covered, and offers closing reflections on how and why location data capture, while having shifted significantly over the course of the past decade or so, will remain a central concern within and for new technological developments.

Having provided this summary of the aims and content of the book, it is time to turn now to Part I, Chapter 1, and an examination of the location-based services ecosystem.

The Topography of Locative Media

Introduction

Part I of the book, entitled "The Topography of Locative Media," includes three chapters that, in combination, develop a detailed picture of the business side of locative media.

Given the proliferation of location-based services (LBS) and devices, and in order to be able to examine the broader cultural economic implications of locative media, it is first necessary, in Chapter 1, "Location Services Ecosystems," to delimit the field by asking, what is it, precisely, that we are referring to when we talk of locative media as species of LBS? And, how might we give shape to and begin to discuss LBS as *an industry*? The argument of this chapter is that, despite the multiplicity and apparent dispersal of LBS, by taking an "ecology of communication" approach (Altheide 1995), it is possible to both define *what* we regard as LBS technologies and *who* the key corporations are who are responsible for their development and operation, and *how* the various companies and the services they provide interact. In the chapter, I concentrate on the mobile location ecosystem and argue that, in its industrial formation, it can be understood as composed of dynamic interactions between location management and content services firms, which primarily interact with platforms and ad monetization firms, which in turn principally interact with various social and location-oriented applications. In detailing these interactions, a clearer picture of the industrial composition of mobile LBS emerges, and it is a picture with "developmental, contingent, and emergent features," which suggests that the ecosystem "does not exist as a thing" necessarily, but rather, given the still emergent nature of LBS, as a "fluid," ever-shifting structure or set of relations (Altheide 1995, 10–11).

Chapter 2, "The Business of Maps," considers the importance of maps and mapping assets within the larger location-based services ecosystem. Maps, in one form or another, are a fundamental aspect of all locative media. Within the available critical literature on mobile communications and maps, a great deal of attention has been given to charting and making sense of the technical, cultural, and phenomenological dimensions of mobile media and map use. However, far less attention has been granted to understanding how maps and map infrastructures are built and maintained, and the commercial importance of and transactional processes driving the control of mapping data. Maps are a vital component of the larger political economy of location-based services. One 2013 estimate placed global geo services revenues at between US$150 billion and $270 billion per year (Oxera Consulting 2013, 1), figures that are likely to have climbed very significantly since then. The rise of smartphones has meant that maps and mapping have come to perform important functions for end-users in, for example, aiding urban navigation and in supporting a variety of social practices, such as mobile social networking and the use of local recommendation services. This chapter examines the business of maps and the cartographic infrastructure that sits behind many location-based mobile services. Specifically, the chapter will consider the arrival of Apple Maps, Apple's efforts in building a maps service to rival Google's, and its longer-term ambitions to extract commercial value from its maps data.

Chapter 3, "Location Integration and Data Markets," examines the geolocation data-capture efforts, and still-evolving business and revenue models, of two commercial businesses now central to the contemporary settlement of locative media. One of these firms is clearly identified as offering location services (Foursquare), while the other was not, until recently, generally thought of as location-based (Facebook). In combination, these two cases provide a rich composite picture of the ecologies of locational information, and how each company, in interaction with the affordances of its specific technologies and patterns of use, generates location data, and how these data are harvested, stored, valued, and commodified.

The arguments that are developed in the three chapters in Part I draw extensively from trade press reportage and other forms of industry commentary (such as blog posts by industry analysts). As reliable corporate data are notoriously difficult to obtain, especially data relating to start-ups and privately owned firms, trade papers in particular remain a vital resource for scholarly research into the recent historical and political economic developments of companies operating within the fast-moving field of communication infrastructures and networked media (Corrigan 2018; Wilkinson and Merle 2013, 417). Thus, critical engagement with trade press sources can assist in building a more complete picture of these arrangements. Consulting a range of trade press and related sources can prove invaluable in reducing "information asymmetry" and other factual inaccuracies, thereby enabling greater depth of analysis through the cross-checking of multiple sources, especially around earnings reports, firm-initiated disclosures, and market reactions to these (Bushee et al. 2010, 1). In the present context, critical engagement with the trade press provides important insight into the rapidly shifting corporate landscape relating to communication infrastructure and location-based services, the business structures of and business decisions made by these firms (Mintz and Schwartz 1985, xvii), as well as everyday representations of these firms.

CHAPTER 1

Location Services Ecosystems

Location-based services will become as common as today's yellow pages.

—Harel Kodesh, president of Wingcast Incorporated, quoted in Auto Channel (2001)

INTRODUCTION

The preceding deceptively simple statement, made in 2001 by a US automotive executive, that compares the rise of location-based services and the apparent ubiquity of telephone directories, points to a number of concerns that are of direct relevance to the discussion of this book as a whole and of this chapter in particular.

In an article examining telephone directories as a species of list media, Gerard Goggin (2012) details the fluctuating contemporary fortunes of these directories and corporate control of them. During the 1970s and 1980s, the monopolistic nature of corporate control of telephone directories (directory ownership was dominated by telecommunications companies, or "telcos"), and "their relative lack of competition, and their significance in advertising for small business in particular," meant that these directories became a target for deregulation (Goggin 2012, n.p.). With "deregulation of telecommunications around the world from the 1980s onwards, the value of telephone directories and directories-associated services" rose quite dramatically (Goggin 2012, n.p.). However, over the ensuing decades, there were successive legal challenges in different jurisdictions that sought to undermine the domination of directory services by telephone carriers. These legal challenges became increasingly centered around the databases that sit at the heart of directories, and, indeed, those firms with stakes in yellow pages had to adjust their commercial interests from operating yellow pages as a form of media to seeking to protect the value of the databases underpinning them. By 2012, telcos were, at best, keeping only a "precarious grip on the heritage of the telephone directory" (Goggin 2012, n.p.).

This adumbrated account of the struggle for control of telephone directory services is instructive here in that it reconfirms just how susceptible to change media industries can be and are, with firms experiencing sometimes seismic shifts in fortune within a relatively short period of technological development. Thus, while figures like Harold Kodesh predicted that location-based services would become commonplace, as yellow pages once were, it is important to bear in mind that what *form* these services are likely to take in the near and distant future, and who is likely to control them, will be subject to ongoing evolution and change. Wingcast Incorporated stands as a case in point. It was formed as a telematics joint venture between Ford Motor Company and semiconductor and telecommunications equipment firm Qualcomm Incorporated. The company was founded in 2000 in order to develop information services (combining voice, entertainment, and internet access) and safety and security features for automobiles using Qualcomm's CDMA (Code Division Multiple Access) wireless technology. However, less than two years after its formation, the joint venture was dissolved by Qualcomm (Fleet Owner 2002). But now, within the space of just over a decade, it seems that we have come almost full circle, with reinvigorated interest in telematics and location technologies from Ford (among a host of other automotive manufacturers and tech companies) due to the promise of the convergence of location-based services and driverless vehicle technologies.

Thus, the comparison that Kodesh draws in the opening epigraph is illuminating in that it points to two, important, interconnected features of the larger industrial composition of location-based services. The first of these features is the dynamism of this industry, which will be discussed in further detail in this chapter. The second feature is the evolving nature of the business and revenue models that are adopted by firms operating within this industry, which will form the focus of successive chapters.

What precisely do we mean when we talk of location-based services (as opposed to locative media, more narrowly defined)? And, how might we give shape to and begin to discuss location-based services *as an industry*? Given that the larger ambition of this book is to develop a detailed portrait of political and cultural economies of locative media and their implications, it is first valuable, given the proliferation of broader location-based services, to answer these questions.

Initial academic presentations on the research informing this book met with repeated questions about whether it was even possible to conceive of location-based services *as an industry* given the scale and diffusion of global geoservices and the proliferation of firms involved in the provision of these services. In 2016, location-based services site Geoawesome made an attempt at identifying the "top 100 geospatial startups and companies in the world" (Buczkowski 2016b). In the comments threads that followed, prominent US-based maps industry analyst Marc Prioleau queried the usefulness of such an endeavor, remarking that "a list of 100 doesn't really narrow anything down" (Prioleau, in reader comments in Buczkowski 2016b); other comments noted "glaring" omissions in the top 100, suggesting an expanded list was necessary.

Despite these research and taxonomical challenges, and despite the complexity and apparent dispersal of location-based services (hereafter shortened to LBS), I propose that, by adopting an ecological or ecosystems approach, it *is* possible to narrow down *what* we regard as LBS technologies, to delineate *who* the key corporate players are that are responsible for their development and their (inter-)operation, and to explain *how* these various companies and the services they provide interconnect and interact.

ADOPTING AN ECOSYSTEMS APPROACH

One of the benefits of the ecological metaphor, Greg Hearn and Marcus Foth (2007) suggest (and what makes it particularly helpful for giving form and shape to a diffuse field such as LBS), is that it "focuses on whole of system interactions," "enabl[ing] us to define the boundaries of any given ecology" and assisting us in examining "how the coherence of that boundary and the stability of each ecology is maintained."

Steward Pickett and Mary Cadenasso (2002, 6) note that the concept of an ecological system, in its metaphorical application, "can be used to stand for equilibrium, resistance or resilience, diversity, and adaptability." Approaching LBS as an ecosystem (or, more accurately, a series of interlocking ecosystems) in all of these various senses can prove instructive for making sense of LBS as an interconnecting set of industries with various internal forces and dynamics.

There are a number of aspects to the preceding understanding of the various senses of the ecological system metaphor that are germane to the discussion to follow. First, an ecology or ecosystem implies "relationships related through process and interaction"—relationships, furthermore, that foster "interdependence, mutuality and co-existence" (Altheide 1994,

667) and "opportunities for successful adaptations" (Nardi and O'Day 1999, 96). This understanding proves helpful, for instance, in this and succeeding chapters for grasping how it is that a location data intelligence firm such as Foursquare can coexist and interact with Alphabet subsidiary Google, which is operating in a similar space, or how the autonomous vehicle operations of Google and Uber can coexist with some degree of interdependence and mutuality (legal battles notwithstanding). These interdependencies lead José van Dijck (2013a), in her study of social media platforms, to refer to them collectively as comprising an "ecosystem of connective media" (18) made up of manifold interacting "chains of microsystems" (163).

Second, within an ecosystem or set of interacting ecosystems, like that comprising LBS, "relations are not haphazard or wholly arbitrary" (Altheide 1994, 667). Rather, "connections have emerged that are fundamental for the medium (technology) [or firm] to exist and operate as it does" (667). Here are three short examples of such connections pertaining to the LBS ecosystem:

(a) Banjo, a social media image analysis firm, has been able to exist and grow in large part as a result of cross-platform access to the image-based social media feeds of other services, such as Instagram, and, it is claimed, Vkontakte and WeChat, among others (see Wilken 2018b).

(b) Neustar's Localeze works with companies to manage the identity and veracity of local web listings information, and continues to operate due to the services it provides to smaller firms (especially local businesses) and the connections and partnerships it has forged with larger search, geo, and social media firms (including Yahoo, Bing, Yellow Pages, TomTom, Apple, Twitter, and Facebook) (see https://www.neustarlocaleze.biz/).

(c) What3Words, which was launched in 2013, pioneered a unique three-word combinatorial system for responding to map addressing issues, and has been able to grow as a result of strong connections both within and outside of the existing LBS ecosystem (see https://what3words.com/).

Third, different species occupy different positions of prominence within an ecological system. For example, Nardi and O'Day (1999) describe how an ecology is often "marked by the presence of certain keynote species whose presence is crucial to the survival of" (53) and to the "robust functioning of the ecosystems of which they are a part" (80). At present, within the LBS ecosystem, Google (and its parent company Alphabet Inc.) are one such keynote species. Yet, as Nardi and O'Day go on to note, other, often smaller, species "take advantage of different ecological niches, which provide natural opportunities to grow and succeed" (51). Just as within an ocean system remoras and other sucker fish take advantage of the opportunities presented by keynote species like dolphins and sharks, so, too, in the LBS system, we see smaller firms, like Button and IFTTT, which provide interoperability-related services, taking advantage of the opportunities presented by large firms like Google and Facebook by facilitating interactions between their platforms, clients, and the online offerings of other, smaller firms and start-ups with which they interact.

Fourth, as Altheide (1994, 667) observes, "there are developmental, contingent and emergent features" of ecological systems; an ecology, he writes, "does not exist as a 'thing,' but is a fluid structure" (667). That is to say, as Nardi and O'Day (1999, 51) put it, "change in an ecology is systemic"—there is always "ecological evolution" (97). What is important about ecosystem change is that its impacts "can be felt throughout the whole system" (51), and any adaptations that arise within the ecosystem as a result "lead in turn to further change, as the entire system adjusts to new constraints and possibilities" (52). As Matthew Fuller (2005, 1) argues, "complex objects such as media systems" should be understood as involving "processes embodied as objects, as elements in a composition," which settle "temporarily into what passes for a stable state" before reforming and resettling, and so on, in a process that is ongoing. For instance, Foursquare, which has long been the incumbent in location-sensitive mobile social networking (LMSN), has had to adapt and change in response to multiple waves of reformation and resettlement that have reshaped this particular subfield of LBS. And, perhaps most infamously, the dramatic and controversial arrival of Uber has led to widespread industry and regulatory upheaval in virtually every national market it has entered. Such rapid and relentless industrial change is why José van Dijck (2013a, 22) refers to the social media industries as an "unstable ecosystem" due to the "volatile" nature of the "larger technology infrastructure."

An appreciation of "ecological evolution" also holds methodological value insofar as it provides a productive means of navigating around one of the key issues attending much digital media research: the speed with which firms and applications tend to come and go, often before a research project is completed (Frith 2015b, 9–10). We see this, for example, in the three generations of LMSN I described in the Introduction that have occurred since the advent of the smartphone. An acceptance of evolutionary change usefully shifts attention away from whether a particular application or firm survives to "remaining focused on specific capabilities and consequences" (Baym 2010, 13).

In adopting an ecosystems approach, the aims of this chapter are to highlight the diversity of LBS; to give form and shape to this ecosystem; to describe some of the constituent "species" (the key corporate players that occupy this ecosystem); and to detail the ways that the different parts of this ecosystem work together. This exploration of LBS ecosystems proceeds in two steps, each at a different scale. In the first step, I give a detailed account of what US maps industry analyst Marc Prioleau (2010) calls the "mobile location ecosystem." Examining this ecosystem is important, as many of the platforms and applications discussed in this book are to be found here. The second step involves broadening the focus so as to view the mobile location ecosystem as one among many other (often much larger) interconnected ecosystems. While the account that is offered here can only ever be indicative rather than exhaustive, it nonetheless serves to circumscribe the overall terrain being explored in the book, and it provides some insight into the many important interactional complexities at play.

MAPPING THE MOBILE LOCATION ECOSYSTEM

To try to make sense of LBS, I begin by drawing on Marc Prioleau's (2010) diagram of the mobile location ecosystem (see Figure 1.1). While intended as a "high-level," indicative schematic representation only, Prioleau's account is nonetheless instructive for grasping key interactions within the mobile LBS ecosystem. Prioleau's suggestion is that, at an industry level, the mobile location ecosystem can be understood as composed of dynamic interactions between different industrial sectors within a larger LBS ecology. These include firms handling various facets of "location management" (that is, technological systems and services used in determining one's position and then managing that location) and "content" firms (that provide geospatial and non-geospatial content in the creation of applications) (Prioleau 2010). Location-management firms primarily interact with "ad monetization firms" (advertising and marketing platforms that supply location-based ads and offers, and analytics platforms that analyze and derive context from location-based data), and content firms primarily interact with geo-focused "platforms" (multi-application platforms that support maps and routes, and/or provide geocoding and a places directory) (Prioleau 2010). These four sectors interact with a fifth: "applications" (which include "multi-thousand participants" across major areas such as navigation, social networks, local search, and commerce) (Prioleau 2010). This last sector—"applications"—holds prime position in these interactions in Prioleau's diagram, presumably for the reason that applications tend, for

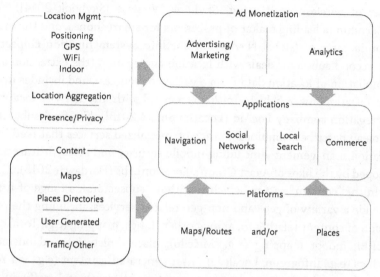

Mobile Location Ecosystem

Location Mgmt

Positioning
GPS
WiFi
Indoor

Location Aggregation

Presence/Privacy

Ad Monetization

Advertising/
Marketing

Analytics

Content

Maps

Places Directories

User Generated

Traffic/Other

Applications

Navigation

Social
Networks

Local
Search

Commerce

Platforms

Maps/Routes   and/or   Places

Figure 1.1. The mobile location ecosystem.
Source: Image courtesy of Prioleau Advisors (https://www.slideshare.net/mprioleau/mobile-location-ecosystem). CC-BY-SA.

now at least, to be the most prominent sector of the LBS industry of interest and concern to end-consumers.

Adding further detail to this portrait, Prioleau suggests that each of these five main sectors is also composed of various subsectors, or "chains of microsystems," as van Dijck (2013a, 163) would call them.

In the following, I provide detailed descriptions of each subsector, or "microsystem," and examples of the sorts of firms that operate within them. This description is warranted for three reasons. First, it is provided for the benefit of those readers not acquainted with the firms and interfirm interactions involved in the LBS industries. Second, it is my hope that through this description, one might acquire a cumulative sense of the breadth and complexity of the corporate involvement in the mobile location ecosystem. And, third, being able to grasp something of this complexity is valuable context for the examination of specific companies that will follow over the next two chapters.

Let us begin describing each of the sectors' respective subsectors (or "microsystems") within Prioleau's diagram, moving left to right, and top to bottom, starting with "location management." "Location management," Prioleau suggests, comprises three subsectors. The first features firms specializing in positioning technologies (GPS, wireless, and indoor), including, for instance, TruePosition (who, in 2014, acquired a key provider of wireless positioning, Skyhook Wireless [Skyhook 2014]) and Qualcomm (a leading maker of processors for smartphones, and the main manufacturer of Global Navigation Satellite System function chipsets). The second subsector deals with location aggregation (the extraction and compilation of location data from a variety of sources), and includes firms like LocationSmart (which, in 2015, merged with another key location-aggregation company LocAid [LocationSmart 2015]). The final subsector Prioleau describes combines a range of specialized services that feed into location management—including mobile security and safety features developed by the likes of Avast Group's LocationLabs (Hardawar 2013).

In the "content" sector (which is without subsectors), a range of firms provide a variety of geo- and non-geo-related services, including the provision of content relating to, among other things, mapping (e.g., TomTom, HERE); indoor mapping (e.g., Micello); place directories and business listings (e.g., Infogroup, Localeze); "user-generated" content (e.g., via Yelp and other social media and recommendation services); and traffic information (e.g., INRIX, a firm founded in 2004 using software licensed exclusively from Microsoft [PRNewswire 2005], which has since grown to become the market leader in the provision of real-time and predictive traffic information).

Within the "ad monetization" sector, firms provide advertising and marketing expertise. The industry heavyweight here is Google's Admob, but there are numerous others, including Placecast, which builds "audience profiles based on behavioral location rules, and other behavioral device data" (Placecast 2017) and PlaceIQ (with one of its investors being e-commerce giant Alibaba [Buczkowski 2016a]). In addition, there are also analytics firms, including location-based social media marketing customization firm MomentFeed (Magistretti 2017), YP [Yellow Pages] (which, in 2014, acquired Sense Networks, a firm that specialized in analysis of large-scale location data for marketing purposes [Ha 2014]), and Hootsuite (which, back in 2001, purchased specialist location marketing support firm Geotoko [Parr 2011]).

The "platforms" sector includes two subsectors. One of these relates to the provision of maps and routes data. The clear industry leader here is of course Google, but there are several other important players, including TomTom, TeleNav, HERE, Microsoft's Bing Maps, AOL's MapQuest, Open Street Maps (OSM), and Uber and Apple, with the last of these to be examined more closely in the next chapter. The other subsector includes rich places databases (and corresponding APIs) that are held by search and social media firms like Google, Facebook, Foursquare, and Twitter, as well as other, less well-known firms, such as Factual (which acquired the places data from the APIs of geo infrastructure firm SimpleGeo, following the latter's acquisition by Urban Airship in 2011 [DuVander 2012]).

And, finally, the "applications" sector comprises a diverse and multitudinous array of services pertaining to navigation (including Google, TomTom, Telenav, Garmin, Sygic, and iGo), social media (such as Facebook and its subsidiaries, Swarm, Snapchat, and Twitter), local search (like Foursquare, Yelp, YP [Yellow Pages], and Zomato), and commerce (from big players like eBay and Alibaba, to smaller players, like the South Korean SKP-owned local shopping rewards service, Shopkick).

Broadly speaking, there are two main ways that these various sectors (and their requisite subsectors) tend to combine and interact. The first follows the model that Prioleau describes as "self-contained location"—which is much like the "full service" approach once dominant within the global advertising industry (Sinclair 2012, 38), where a parent company has specific arms or subsidiaries that fulfill a variety of functions that combine to offer a full range of services to clients. Alphabet's Google is a prime example of a company—or, more properly, suite of companies—that seeks to offer close-to-full service, or "self-contained location." This was perhaps most clearly evident, circa 2010, when Google possessed distinct offerings across all the sectors captured in Prioleau's diagram—for example, within location management (Google Latitude), content (Google Street View, etc.), ad monetization (AdMob, AdSense), platforms (Google Maps, Places), and applications (Google Maps Navigation, Orkut, Google Local, Google Checkout). While Google's specific location offerings are perhaps more opaque since the establishment of its holding company structure under Alphabet Inc., its "full service" ambitions nonetheless remain strong, and, indeed, its ability to provide "self-contained location" services is arguably even more pronounced under its new corporate arrangements, especially given its continued investments in location technologies—a point that will be returned to later in this chapter.

The second, and more prevalent, approach is a partnership model, which sees smaller firms, in particular, partnering with LBS specialists to provide products and services that ultimately feed into consumer-facing applications. This has been a long-standing strategy. Back in 2002, for instance, mobile telecommunications company Hutchison 3G UK announced an agreement with five companies—whereonearth (since bought by Yahoo) in partnership with Telcontar (which became deCarta and has since been acquired by Uber), the UK Ordnance Survey, TeleAtlas (a wholly owned subsidiary of TomTom), IONIC Software (an operating system software design kit [SDK] for mobile app development), and TCS (Tata Consultancy

Services)—to deliver location-based services to Hutchison customers. These services would permit Hutchison customers to "pinpoint their position on a digital map delivered to [a] 3G handset, then follow directions to services such as hotels, restaurants, shops and businesses using a detailed real-time display" (Hutchison Whampoa Limited 2002, n.p.). This early partnership is significant in revealing the non-trivial nature of establishing mobile geolocational operability. Establishing location-sensitive mobile marketing efforts required collaboration between multiple companies with a range of specialist skills and expertise.

Beyond this early example, partnerships remain the prevalent model within the mobile location ecosystem. The case that Prioleau cites as illustration of the partnership approach, which dates from late 2010, shows how internet of things (IoT) application firm Fullpower formed partnerships with Skyhook Wireless and Google Latitude, Navteq, and deCarta for the provision, respectively, of location management, content, and platform-related maps data (Prioleau 2010). Furthermore, due to the ongoing consolidation of geo-related resources and services by Google, partnerships remain a necessity for other firms if they are to resist Google's stranglehold on mobile LBS. For example, Apple, as we will see in the next chapter, has partnered with a range of firms while it pursues its longer-term goal of building its own maps-related products and services, and thus freeing itself from reliance on Google's LBS offerings.

There are two final observations to make about Prioleau's diagram. First, and to reiterate a point made earlier in this chapter, ecological evolution is constant, and, year after year, LBS firms' fortunes wax and wane, and the firms themselves come and go, often at an astonishingly rapid rate. Indeed, to return to the last example provided in the preceding discussion, while IoT firm Fullpower still operates, all the partner firms have undergone significant change: Latitude was closed by Google in 2013; Skyhook Wireless, as noted previously, was acquired by TruePosition in 2014; Navteq fully merged with Nokia in 2011, becoming Nokia HERE, which in turn was purchased in 2015 by a consortium of three German car manufacturers, BMW, Daimler, and Audi; and deCarta was sold to Uber in 2015 (see Wilken and Thomas 2019). Given that industry tumult of this sort is to be expected (ecosystems are dynamic systems), what has been of principal interest are the types of interactions occurring within this system. Second, firms that started out operating in one subsector quite often expand to operate in more than one, and/or shift operational focus, moving from one subsector to another. A case in point, and one that will be explored in more detail in Chapter 3, is Foursquare, which spun off the social media aspects of its service into its Swarm app, transformed its Foursquare app to become a search and recommendation service, all the while strengthening its places database and building up its enterprise offerings.

In this first part of the chapter, I have described Marc Prioleau's diagrammatic summary of the mobile LBS ecosystem, and the interactions between its various subsystems, in some detail. As noted at the outset, I have dwelt on this diagram at length for the reason that it provides a productive point of entry into, and a valuable snapshot of, the structural and interactional complexities of this ecosystem. Moreover, it also delimits what is the main focus of this book: mobile LBS, or "locative media," as it will be referred to in the chapters that follow. Indeed, for the most part, the focus of the book is narrower still, in that it is largely restricted to a discussion of locative media "platforms" and "applications" (in Prioleau's sense of these terms).

While mobile locative media platforms and applications form the main focus in later chapters, it is important to note that the preceding discussion, based on Prioleau's diagram, is just one rendering of the mobile location ecosystem, and one that will always be indicative rather than exhaustive. What is more, this ecosystem continues to evolve. Prioleau published his diagram in 2010. In 2016, geoservices website Geoawesome published its own system of classification for grouping contemporary geo-spatial related firms. This classificatory system includes many of the kinds of firms already discussed, as well as a range of business categories, such as "drones" and "location big data," that have only recently risen to prominence.

In the next chapter section, I wish to explore how the mobile location services ecosystem, which Prioleau's diagram attempts to capture and describe, overlaps and interacts with a number of other, equally significant infrastructure-related ecosystems. What follows is a macro-level outline only, and is not intended to offer detailed critical analysis of each larger infrastructural system.

"SCALAR ENTANGLEMENTS" OF INFRASTRUCTURAL ECOSYSTEMS

Writing on communicative ecologies, Jo Tacchi (2006) argues that it is important to remember differences of context and scale. Drawing on extensive fieldwork in the area of information and communication technologies (ICTs) for development, Tacchi (2006) argues that, in the study of communication, the ecology framework is useful insofar as it focuses attention not just on more immediate communication-related aspects of the contexts in which people operate, but also on the ways that they are "in turn imbricated in other structural, social, economic and cultural contexts." The various imbrications—or "scalar entanglements" (Taffel 2013)—that Tacchi describes here are valuable to bear in mind when thinking about the interactions between the mobile location ecosystem (described earlier) and larger infrastructural ecosystems (described in the following).

What is also valuable about ecological or ecosystem metaphors is that they fruitfully capture "the massive and dynamic interrelation of processes and objects, beings and things, patterns and matter" (Fuller 2005, 2). The emphasis that Matthew Fuller gives to objects here is especially important in that it points toward more recent work on communicative and media ecologies that underlines the need to "rethink the material basis of [the] contemporary media condition" so as to produce "more complex intuitions that take into account a certain 'activity of matter'" (Parikka 2010, xx).

One influential strand of such work that has emerged from a reinvigorated theoretical interest in materiality, and which informs a broader understanding of LBS, can be understood through the rubric of "new materialism" (Coole and Frost 2010b; Dolphijn and van der Tuin 2012). "New materialism" advances the claim that "foregrounding material factors and reconfiguring our very understanding of matter are prerequisites for any plausible account of coexistence and its conditions in the twenty-first century" (Coole and Frost 2010a, 2). As Jeremy Packer and Stephen Crofts Wiley point out, this has led to work (re)focusing on what they term the "materiality of communication"—that is, "communication infrastructure, transportation and mobility, mobile technologies, and the production of urban, regional, and translocal spaces" (Packer and Crofts Wiley 2012, 12; see also Gillespie, Boczkowski, and Foot 2014). The "materiality of communication" provides a particularly productive frame, or point of entry, for thinking about infrastructure. A focus on infrastructure is important, for, as Susan Leigh Star (1999, 378) notes, "the study of information systems implicitly involves the study of infrastructure."

Infrastructure is here understood as a "system of substrates" ("railroad lines, pipes and plumbing, electrical power plants, and wires") that enables other things to happen (Star 1999, 380). Brian Larkin (2013, 329) argues that, at an even more fundamental level, infrastructure can be understood as "matter that enable the movement of other matter." This is to say that infrastructures "are things and also the relation between things" (329); these "things" might be "built things, knowledge things, or people things" (329). As Paul Edwards (2003, 185) puts it, infrastructures "are the connective tissues and the circulatory systems of modernity." In what follows, four larger infrastructural systems can be thought of in the preceding terms. Each, it should be noted, is by no means tightly defined (there is considerable "infrastructural heterogeneity" [Vertesi 2014, 267]) or neatly contained (all involve "multi-infrastructural interactions" [266]). I should also note that, as with the preceding discussion of mobile LBS, the intention here is not to provide a comprehensive account, but rather to form an outline of four, large infrastructural systems with which mobile location services interact and on which they rely. These four sets of infrastructures I am labeling as follows: *media infrastructures; geospatial data infrastructures; coordination and control infrastructures*; and *finance infrastructures*.

Media Infrastructures

The rubric of *media infrastructures*, as it is conceived by Lisa Parks and Nicole Starosielski (2015a), is an expansive category that incorporates a range of telecommunications, mobile, and computing-related technologies and other systems supporting mobile location–related services.

If we begin with the mobile handset itself, it can be regarded as a vital enabling "substrate" of mobile location services. These devices are the end result of "three entangled relational scales of content, software, and hardware" (Taffel 2013), each with its own complicated trajectories of commercial design and manufacture (see, for example, Goggin 2006; Parikka 2015; Qiu 2016).

In the case of smartphones, additional infrastructural layers also support the availability of mobile applications (apps). Crucial to understanding the mobile app economy is recognizing that it is, in reality, "a collection of interlocking innovative ecosystems," and that at the heart of each ecosystem are three things: "a core company, which creates and maintains a platform, and an app marketplace, plus small and large companies that produce apps and/or mobile devices for that platform" (Mandel, quoted in Goldsmith 2014, 171; see also Van Alstyne et al. 2016; Wilken 2018a).

What is more, our use of mobile handsets—and the location services enabled by and through them—also rely heavily upon other, larger media infrastructures. As Nicole Starosielski (2015a, 53) points out, "with each wave of technological development, the media landscape appears less wired." And, yet, wireless and wired technologies are also often more closely associated than we tend to recognize and acknowledge (see Farman 2015d). For instance, the "proliferation of wireless media technologies," Starosielski (2015a, 54) writes, "is grounded by a large mass of [underground and undersea] cable systems" such that "every time users search for information using Google, post a picture to Instagram, or dial a number on Skype, they activate a part of this subterranean and subaquatic infrastructure" (see Figure 1.2).

Undersea cables, though, are merely the "deepest strata" (Starosielski 2015a, 55) among many layers of "vertical mediation" (Parks 2015, 2016, 2017; see Starosielski 2015b). Starosielski's (2015a) point about the "grounding" of internet use in undersea communications infrastructures also applies, of course, to manifold above-ground infrastructural systems, including cell towers, telecommunications backhaul networks (that include fiber optic, coaxial, and hybrid forms of cabling), mesh networks, and Wi-Fi hotspots and routers, which support our use of location-enabled mobile devices that combine to create "signal territories" (Parks 2013).

The dramatic scaling up in the volume and complexity of "signal traffic" (Parks and Starosielski 2015b; see also Dourish 2015), has also given rise to other, even more recent forms of media infrastructure: namely, international data centers associated with cloud computing and the "storage, processing, and distribution of data, applications, and services for individuals and organizations" (Mosco 2014, 17). Data centers function, in essence, as distributed systems that "store content out on the network rather than on

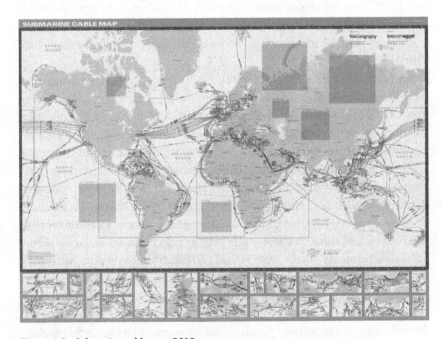

Figure 1.2. Submarine cable map 2018.
Source: Created by Markus Krisetya, Larry Lairson, and Alan Mauldin. © 2018, TeleGeography, a Division of PriMetrica Inc. Reproduced with permission.

a personal computer" or portable device (Starosielski 2015a, 53), a capacity that is claimed to be crucial for the successful real-time functioning of a range of location-related processing and analytics services (for critical discussion, see Holt and Vonderau 2015; Rossiter 2016, 2017).

It should also be noted that these contemporary media infrastructures, which facilitate the sorts of "trafficked signals" (Mattern 2015, 94) crucial to the successful operation of mobile location services, form part of a much longer history of media infrastructural systems—what has been referred to elsewhere as the "deep time of media infrastructure" (Mattern 2015; Zielinski 2006; see also Bollmer 2016; Farman 2018b; Mattern 2017; Parikka 2012).

Geospatial Data Infrastructures

A second, larger system the mobile location ecosystem relies upon and interacts with is that of *geospatial data infrastructures* (Coleman and McLaughlin 1998). I am using this as an umbrella term for capturing a range of geo-related services and supports that relate to maps and mapping and other forms of geographical information systems (GIS), and satellites.

With respect to the last of these, Parks and Starosielski (2015a) discuss satellites as forms of media infrastructure; I discuss them as part of geospatial data infrastructures for the reason that I am less focused on their use within telecommunications (Aked 1990), or in the transmission of broadcast content (see Aslinger 2012; Morley and Robbins 1995; O'Neill and Murphy 2012; Parks 2005; Russo and Kirkpatrick 2012; Sakr 2012; Torre 2012). Important as these applications are, my interest in this book is on their positioning and remote sensing capacities as "Earth observation media" (Packer and Reeves 2013), and their ability to "turn vast swaths of the Earth's surface into signal territories" (Parks 2013, 286). It is these capabilities that lead Lisa Parks (2012, 133) to describe satellites as forms of "meta-infrastructures" in that they can, in certain cases, extend "transcontinental footprints, thereby interlinking and leapfrogging some terrestrial infrastructures."

The satellite industry has its own particular histories of development (Sturdevant 2012), both within and increasingly outside of "Western" military-industrial contexts (Aslinger 2012; Erickson 2012; Parks 2012; Sakr 2012), its own regulatory regimes (Collis 2012), and its own industrial composition that, like all the ecosystems described in this chapter, is subject to evolutionary change (Foust 2017). This is also the case, of course, for the other elements that form the geospatial data infrastructures ecosystem, digital mapping (e.g., Coast 2015; Kilday 2018) and GIS (e.g.,

Leszczynski 2012; Wilson 2017). In noting these developments here, my point is simply to emphasize that when mobile location services interact with geospatial data services, these interactions may appear seamless to us as end-users of mobile phones (indeed, "so central to daily life in advanced capitalist societies as to become invisible, part of our technological unconscious" [Wilson 2016, 288]), yet they are in fact the result of complicated intra- and inter-ecosystem interactions.

Coordination and Control Infrastructures

A third set of services upon which LBS relies, and with which LBS interacts, are *coordination and control infrastructures*. In developing this particular categorization, I follow Bowker et al.'s (2010, 98) understanding of infrastructures as not only "tubes and wires" and technologies, but also as including a variety of supporting organizations, agencies and societies. Drawing from the work of Jan van Dijk (2012, 74), I conceive of coordination and control here as different means of managing complex industrial, bureaucratic, regulatory, and policy processes. *Coordination* involves the systematization and synchronization of technical processes that are often developed within and managed by scientific agencies, and professional societies and associations (like the IEEE or the Wi-Fi Alliance), and which are implemented through the creation of, for example, technical formats and standards (such as mobile phone standards, or the IEEE 802.11 suite of standards, which include a set of specifications for implementing wireless local area network [WLAN] computer communication in a range of frequencies) (see Goggin 2006; Mackenzie 2010). *Control* involves attempts at management by command and authority, and involves policy regimes (see Braman 2009; Iosifidis 2011), regulatory measures (most notably, radio frequency spectrum allocation), and a wide variety of legal frameworks. These forms of coordination and control can be thought of in infrastructural terms insofar as they adhere to Star's (1999, 380) understanding of an infrastructure as (among other things) a system that both enables and constrains. Mobile location services are reliant on these coordination and control infrastructures for their successful operation. There are, however, also times when corporate players and government agencies operating within, or with a strong vested interest in, the mobile location services ecosystem seek to resist, circumvent, or ignore these coordination and control measures (see Chapters 7 and 8).

Financial Infrastructures

The final larger system I wish to draw attention to here involves *financial infrastructures*. These are infrastructures (again in Star's and Bowker et al.'s expanded sense of this term) that facilitate the development and growth of mobile location services and the tech sector more broadly.

Given that so many of the mobile location services discussed in this chapter and book have their roots in the US Silicon Valley tech sector, arguably the best-known and most widely discussed of these financial infrastructures involves venture capital investment (Zook 2002). For many US-based start-ups, funding tends to be scaled and, for the smallest, can involve "pre-seed" investment—"the fools, friends, and families," as Fabrice Grinda (pers. comm. 2013) terms them—followed by more formalized, staged funding rounds involving angel investment (Wong 2010) and then, later, venture capital and other forms of investment from a range of funding actors (private equity, growth equity, venture debt, hedge funds), and then via wider financial markets (Mayer 2004, 40; Haislip 2011).

One noteworthy feature of US-focused venture capital investment is its cyclical nature. This cycle begins with the raising of a venture fund. The cycle then "proceeds through the investing in, monitoring of, and adding value to firms; continues as the venture capitalist exists successful deals; and renews itself with the venture capitalist raising additional funds" (Gompers and Lerner 2004, 3). The cyclical nature of capital-raising described here is noteworthy in the context of this chapter in that it draws out the point that the entire Silicon Valley tech industry funding system is structurally dependent on "ecological evolution"—fundraising, investment deals, exits and returns, renewal/raising of additional funds, and so on.

Given the central place that venture capital investment seems to hold within the US tech imaginary and discourse, it might be tempting to think of venture capital investment as the primary financial driver of the (US) tech sector and of mobile location services. However, Mariana Mazzucato (2013) makes the important observation that the state has in fact played and continues to play a vital role in promoting and supporting innovation. For example, tech industry heavyweight Apple is described as having been "able to ride the wave of massive State investments in the 'revolutionary' technologies that underpinned the iPhone and the iPad" (Mazzucato 2013, 88), including GPS (106ff) and a host of "semiconductor devices" (95). In addition to state-sponsored tech innovation, there is also growing critical awareness of the vital (if more diffuse and at times problematic) role that is played by "informal media economies" (Lobato and Thomas 2015) in driving industrial development and innovation.

INFRASTRUCTURAL INTERACTIONS AND THE CHALLENGE
FOR POLITICAL CULTURAL ECONOMY

In the preceding section, I provided brief accounts of four key, larger infrastructural systems. They offer sketched outlines rather than detailed portraits of each system. Nevertheless, they serve their intended purpose of illustrating how the mobile location system, while in itself rich and complicated, cannot and does not exist or operate in isolation, relying on interactions and negotiations across large and equally rich and complicated "infrastructural environments" (Vertesi 2014, 267).

The extent, and significance, of "infrastructural heterogeneity" (Vertesi 2014, 267) and, particularly, of "multi-infrastructural interactions" (266) can't be overstated. These infrastructural interactions proliferate, operating at a range of scales and in a wide variety of forms, as I will illustrate in the following three examples.

The first relates to the "geoweb," which, as Agnieszka Leszczynski (2012, 74) describes it, "consists largely of information 'volunteered' or generated by lay cartographers over the Web." The operation of the geoweb provides a clear illustration of multi-infrastructural interactions. As Sarah Elwood and Agnieszka Leszczynski explain:

> The geoweb consists of hardware (mobile devices), software objects (applications and services) and programming techniques (such as "mashing up" content) that include virtual globes, interactive mapping platforms, spatial application programming interfaces (APIs), and technical standards (such as GPX) that guide its curation, aggregation, and dissemination. (Elwood and Leszczynski 2011, 6)

Here we see the combination of media, geospatial, and coordination and control infrastructures to create this geolocation-related service.

Second, a key feature of the mobile location services ecosystem, as we shall see in Chapter 3, is the proliferation of cross-platforming deals and data-sharing arrangements—such as between Foursquare and Snapchat, or between Facebook and Chinese technology firm Huawei. In addition, there are a host of smaller firms, such as ITTT and Button (as noted earlier in this chapter), whose primary role is to facilitate interoperability and data flow between larger firms and social media platforms, and others, such as Mapbox, Factual, and Stamen, who provide cartographic and design support to locative media services. José van Dijck (2013a, 163–164) productively refers to such arrangements as "chains of microsystems," with the sorts of firms noted earlier serving as the links in these chains, the "interstices between platforms" (156) and other larger companies. These smaller, interstitial firms fit Joshua Braun's (2015, 77) description of "transparent intermediaries," companies whose success is measured by their invisibility to audiences (76) and end-users of the services they facilitate.

The third and final example concerns the emergence of holding company structures among major search and social media firms that seek to maximize the opportunities presented by multi-infrastructural and cross-ecosystem interactions. Key firms that have undergone corporate consolidation and rearrangement of this sort include the London-based Mail.ru Group (see Wilken 2018a), Facebook (see Chapter 3), and Alphabet Inc., the parent company of Google. It is the last of these that I wish to touch on here. Alphabet Inc. was formed through a restructuring of Google in 2015. What is striking about Google, as noted earlier in the chapter, is that, by 2010, it already had services operating across all of the sectors in Marc Prioleau's mobile location services ecosystem diagram (see Figure 1.1). With the formation of Alphabet Inc., it now has subsidiaries operating not just within the mobile location services ecosystem, but also within three out of the four larger infrastructural systems described in this chapter: media (Google, X), geospatial data (Google, Access & Energy, X, Sidewalk Labs), and finance (GV, CapitalG). Not only does this assortment of subsidiaries tighten Alphabet's grip on the provision of "self-contained location," it also strengthens its ability to operate as a conglomerate that can provide "full service" to its clients, much as global advertising holding companies sought to do through the adoption of similar corporate structures (Sinclair 2012; Wilken 2018a), making them a true corporate behemoth of location-based services.

The sorts of corporate arrangements described above present a great number of challenges to media and communication and internet researchers. As José van Dijck and Thomas Poell (2013, 2) point out, although "the underlying principles, tactics, and strategies" of social and locative media platforms and related firms are generally apparent, the intricate connections between platforms and the internal business underpinnings of these connections and other economic arrangements are more difficult to chart. The "toolbox of traditional media industry analysis," Jennifer Holt and Patrick Vonderau (2015, 76) write, becomes "notoriously difficult to apply." Holt and Vonderau argue that, while long-standing research issues (e.g., "concentration of ownership, subsidiaries and tax breaks, operating efficiencies, and industry resources," etc.) remain "useful categories for political economic analysis," large-scale "media infrastructure [and related] industries are analytically distinct from traditional media industries" (78), and require different approaches. In response, Paul Edwards (2003) calls for a multi-scalar approach to the analysis of infrastructural systems. The level of detail required to do justice to the forms of multi-scalar analysis Edwards is calling for is, however, beyond the scope of this book. Holt and Vonderau (2015, 78), meanwhile, call for a "case-based approach" rather than "one-size-fits-all" approach to political economic analysis. What a case-based approach affords is depth and detail of analysis, and the opportunity for careful, close comparison. This is the approach that will be adopted in the two chapters to follow, where I develop a comparative analysis of two different sets of mobile location services firms.

CONCLUSION

In this chapter, I have sought to give shape to, and attempt to describe, location-based services as an industry. In setting out to do this, I have combined detailed description of the mobile location ecosystem with an outline of the larger infrastructural systems that this ecosystem intersects and interacts with. In detailing these interactions, a clearer picture of the industrial composition of location-based services emerges. It is a picture that contains "developmental, contingent, and emergent features," which suggests that the mobile location services ecosystem does not exist as a fixed thing necessarily, but rather, given the still emergent nature of LBS, as a "fluid," ever-shifting structure or set of relations (Altheide 1995, 10–11). In the two chapters that follow, I explore these shifting structures and relations by examining two different sets of firms that are operating within different parts of the location-services ecosystem. In Chapter 2, I chart the efforts of Apple in seeking to build their own mapping capabilities. And, in Chapter 3, I turn to Foursquare and Facebook and an examination of how each set about building location intelligence capabilities.

CHAPTER 2
The Business of Maps

Maps work by serving interests.

—D. Wood (1992, 4)

[I]t looks as if maps and mapmaking [. . .] are set to get more and not less important. What those maps look like and in whose service they are deployed, however, are unresolved questions.

—Crampton (2009, 91)

INTRODUCTION

Of the many and varied ways of registering, representing, and accessing geolocation information, one of the most ubiquitous involves our use of digital and mobile maps. These are now part and parcel of nearly every facet of our daily lives. We navigate using digital, mobile maps, which reveal where we are, where we want to be, and how to get there. We also draw on manifold other services that use maps data to pinpoint our locations. We rely on them to have groceries and other goods delivered to our doors— indeed, the logistics industries and parallel services, especially taxis and ride-sharing, are heavily dependent on them. We make detailed reference to digital maps (often embedded within booking sites) when planning and arranging distant travel and accommodation, and in finding our way to places that are closer to home yet unfamiliar to us. "Mapping is particularly critical to mobile device use" (McQuire 2016, 81). On our mobile phones, "the map app is the new search box" (Waite 2018). We access maps, via mobile apps like Foursquare, Yelp, or Zomato, when planning a night out, and in selecting or recommending a place to eat. And maps-related location data are often embedded in the information we post to social media sites like Instagram, Facebook, and Pinterest. Maps, of course, are not just of value to us as individuals and to industry, maps are also a source of crucial information for intelligence agencies and the military, as well as within crisis communications for facilitating humanitarian efforts during times of conflict and natural disaster.

Given the central importance of maps to such a diverse array of location-based services and scenarios of use, it is no surprise that accurate maps data are a precious commodity, and corporate control of these assets is contested keenly. In this chapter, I build on prior work on the political economy of location-based services (Leszczynski 2012) to examine the business of mobile maps, asking the following questions: Who controls maps data? What are these data? Where do these data come from? What is their quality? What does it take to build new mobile maps? What are the motivations for wanting to build new maps? And what are the business and revenue models associated with these maps?

In exploring these questions, one obvious approach would be to focus on Google, which is the clear industry leader and a corporate behemoth of geolocation services. However, in exploring the business of maps in this chapter, I have chosen not to look specifically at Google. The decision not to focus on Google was made partly because the story of Google Maps has been told in detail elsewhere (Hutcheon 2015; Kilday 2018; see also Mike Dobson's "Exploring Local" blog: http://blog.telemapics.com/). I also felt that the industry dominance of Google Maps, the importance of what they have achieved in the field of digital and mobile mapping, and the level of influence they exert over the field can be revealed in other, potentially richer ways, such as by focusing on the work of their competitors.

Thus, in this chapter I examine the efforts of one of Google's key rival firms—Apple—and its struggles to build mapping capacity of its own at sufficient quality to be able to lessen (if not entirely break from) its reliance on Google. Apple presents an interesting case in that, as is well known, it is a major player in other areas of the mobile location services ecosystem, yet took industry pundits by surprise when it announced Apple Maps in 2012.

Examining Apple is valuable in order to understand the commercial value of maps and geocoded data, and the significance of Google within the larger location-based services ecosystem. Moreover, it is in looking closely at corporate and talent acquisitions and technological developments, in tandem with consideration of larger corporate ambitions, strategic objectives, and likely future directions, that we are able to come to understand why these firms have been and are prepared to invest so much time, effort, and money in maps and maps-related data.

One aspect of my approach that is worth mentioning here is that, in this (and the following) chapter, I present the corporate developments of the firms under discussion in a linear narrative sequence. I appreciate that corporate strategy and technological innovation naturally proceed according to more complicated and messier pathways of development than a linear narrative might suggest—as evidenced, for example, in Bill Kilday's (2018) insider account of the formation of Google Maps. Nevertheless, a sequential approach has been adopted primarily for the sake of clarity of argument.

Before I turn to examine Apple's mapping efforts in detail, it is valuable to provide some sense of the general contours of the corporate maps landscape prior to revelations about Apple's mapping ambitions.

THE DIGITAL MAPS CORPORATE LANDSCAPE, BG (BEFORE GOOGLE) AND AG (AFTER GOOGLE)

In the years immediately prior to the launch of Google Maps in 2005, the location services ecosystem was dominated by a small yet significant number of "keynote species," to use the language of the previous chapter, that were responsible for the provision of maps-related services. (The field of global mapping-related services is vast, and it is not possible to cover this field comprehensively. Thus, in developing this overview, I am referencing only key commercial firms, and am leaving out of the discussion major government and military suppliers of maps-related data, and the innumerable smaller firms and start-ups—except, that is, where they form part of the narrative I am telling about larger firms, including Apple.)

Pre-Google, maps-related "keynote species" included the following firms:

- Navteq, which was founded in California in 1985, and, by the 1990s, generated most of its revenue from licensing its maps database;
- TeleAtlas, which was formed in the Netherlands in 1984, and provided digital maps and maps data for a range of location-based services;
- Telcontar, which was founded in California in 1996, provided a "geospatial platform that, once customized, provide[d] fast answers to large numbers of consumer oriented queries" (Schutzberg 2005);
- MapQuest, founded in the US as Cartographic Services, and provider of free online maps (MapQuest was subsequently acquired by AOL in 1999 for US$1.1 billion, and then by Oath Inc., a subsidiary of Verizon, in 2015);

- Vicinity, a California-based provider of online maps and enterprise location-related services that was founded in 1995;
- Multimap, a UK-based company founded around 1997, that provided street maps, directions, and other location-related information from a public website; and,
- Esri (short for Environmental Systems Research Institute), a geographic information systems (GIS) firm founded in 1969, and perhaps best known as the producer of ArcGIS software products.

It is also important to note the conception of Open Street Maps (OSM) in 2004 (Coast 2015). OSM, a "wiki-based, crowd- and open-sourced mapping organisation" (Gerlach 2018, 27), was established "as a practical solution to the lack of freely available map data" (Coast 2011), and has since come to play a crucial role in relief efforts during times of humanitarian crisis.

Around this time, the big tech firms of the day also began to get into the mapping act. First, Microsoft (MSN) launched MapPoint 2000 in 1999, a commercial services-oriented location software, that subsequently became a web-based service in 2002. In 2002, Microsoft also acquired Vicinity Corporation for US$96 million (Microsoft 2002). Second, Yahoo! launched its own maps service, Yahoo! Maps in 2002. In doing so, Yahoo! dropped former provider MapQuest (which, as noted earlier, was owned by Yahoo!'s then principal rival, AOL), creating its own service in partnership with data providers Navteq and Geographic Data Technologies (GDT, which was acquired by Tele Atlas two years later), and software application providers Telcontar and Sagent Technology (Francica 2002).

With respect to mobile-related services, a key firm was TeleNav, which was founded in California in 1999 to provide GPS-based navigation software on mobile devices. TeleNav has since grown to become a corporation providing a range of geolocation-related services, including GPS satellite navigation, automotive navigation, local search, mobile advertising, and geo-related enterprise services. In addition to TeleNav, many mobile mapping services at this time—which, it should be remembered, was pre-smartphones—were focused around the portable navigation device (PND) market. This was a market dominated by three firms: Amsterdam-based TomTom, which was founded in 1991; Garmin, formed in 1989; and, the less well-known Etak (B. Edwards 2015), which was founded in California in 1983, and which subsequently passed through a succession of ownership changes, starting with News Corp in 1999, Sony in 1996, and then TeleAtlas in 2000.

A key moment came with the launch in 2005 of Google Maps (under the banner of Google Local). Google built its maps around two company and company talent acquisitions ("acquihires"): Where2Tech, a small Sydney-based start-up formed by brothers Jens and Lars Rasmussen, Noel Gordon, and Stephen Ma; and a more established California-based start-up, Keyhole, that was founded by John Hanke, which already had a solid client base and a team of 29 people (Kilday 2018).

Kendra Mayfield (2002), in a review for *Wired* of what she viewed as the parlous state of online mapping at the time, particularly by the then field-leading provider MapQuest, detailed a litany of problems, ranging from substandard routing and directional information, poor addressing, and points-of-interest (POI) data that were slow to be updated. In discussing Google Maps, one needs to be cautious not to overly lionize Google, casting it as a trailblazer that has achieved things with maps that no other firm has managed to do (on this, see Dobson 2012c). Nevertheless, what Google was able to achieve was to rapidly establish a competitive advantage in online and mobile mapping. This was due to a combination of factors. First, populating and maintaining precision maps is an extremely costly exercise (Dobson 2015b; Gale 2012)—indeed, an analysis in 2012 of the cost of maintaining maps suggested an investment of between US$1 billion to US$2 billion per year (Dediu 2015), with the figure likely to be considerably more today. Here Google Maps had a major advantage over many of its competitors in that it had significant financial backing (by the time of the Keyhole purchase, Google had had its IPO, and its AdWords-generated revenue figures were climbing exponentially). The Maps team also enjoyed strong support from Google's founders, Sergey Brin and Larry Page, who both shared an ambitious vision of what maps could add to the firm's search offerings. Second, Google Maps responded to many of the issues identified in Mayfield's state of the field review: Maps had high-quality base map data, with strong points-of-interest (POI) listings; it was fast-loading and quick to refresh, had powerful search capacity, and, vitally for its long-term sustainability and growth, had clear ad-generated revenue growth potential, particularly around local advertising (see Figure 2.1) (Kilday 2018). This is not to say that Google Maps were or are free from data quality issues—far from it (see, for example, Dobson 2014, 2015a). What Google has, however, is the financial means to continually improve its data.

It is no exaggeration to say that the arrival of Google Maps triggered seismic upheavals across the maps industry, changing the corporate mapping landscape dramatically. For the reasons outlined in the preceding discussion, Google rapidly outpaced those firms that had previously held the dominant positions in online mapping and dependent industries.

Figure 2.1. Combining Maps, search, and online advertising, Google has become market leader in search and local advertising.
Photograph: author.

What followed in the two years subsequent to the launch of Google Maps was significant industry consolidation. For example, Microsoft completed a sequence of acquisitions between 2005 and 2007, including remote sensing imagining technology firm Vexcel (Microsoft 2006; Schutzberg and Reid 2006), geolocation services company GeoTango (Schutzberg 2006), and Multimap (Schonfeld 2007).

It should be noted that this wave of consolidation was not just in response to the arrival of Google; it also reflected the broader industry embrace of what is referred to in the social sciences as the "mobility turn" (Urry 2007; Sheller 2017). For instance, Navteq, which just prior to the launch of Google Maps raised US$880 million from its own listing on the stock exchange, was purchased in 2007 by mobile handset maker Nokia for a then staggering US$8.1 billion (Malik 2007), and subsequently integrated into Nokia's maps, which were later rebranded as Nokia HERE. (Nokia later sold HERE to a consortium of German car makers—BMW Group, Audi AG, and Daimler AG [Mercedes]—for €2.8 billion [Sawers 2015].) In 2007, PND maker TomTom fought off Garmin to acquire TeleAtlas for US$4.2 billion (€2 billion) (Palmer 2007). Neither of these last two deals was considered to have helped Navteq or TeleAtlas keep pace with Google: Navteq's revenue stream soon began to dry up post-acquisition, as Nokia had been a key client (Dobson 2011); and TeleAtlas has been described as "a depreciating asset" under TomTom (Marc Prioleau, pers. comm., 2013). It is also worth mentioning here that Telcontar, which rebranded itself as deCarta in 2006, was sold almost a decade later, in early 2015, to Uber for an undisclosed sum (M. O'Brien 2015).

Throughout this industry upheaval, Google continued to strengthen its position. It did this through further acquisitions, including Foursquare predecessor and SMS-based social location check-in service Dodgeball (Block 2005), geo-tagging photo service Panoramio (Hanke 2007), restaurant ratings firm Zagat (Mayer 2011), crowd-sourced traffic information service Waze (McClendon 2013), and Skybox Imaging (Alphabet Inc. 2014), among many others. Google also strengthened its position through ongoing research and development and innovation (R&D+I) efforts, and year-on-year new product releases: Google Earth appeared soon after Google Maps in 2005; mobile web search, an xhtml precursor to what would become the Google mobile app, was announced in 2005; its Maps APIs were opened up in 2005; Local Business Ads with AdWords was added in 2006; Google Street View was controversially rolled out in 2007 (the focus of the first half of Chapter 7); Google Map Maker, which allowed users to submit maps for regions with little or no map data, was also announced in 2008; an Adsense layer was added to its Maps API, so that third parties could earn money by embedding ad-laden maps on their website or in their apps, appeared in 2009; its own stand-alone Maps app for iOS appeared in 2012; and the list goes on and on. As maps industry analyst Marc Prioleau remarks, "every three months they've done new things to get up to this level" (Prioleau, pers. comm. 2013; see also G. Miller 2014). In 2008 alone, Google made 21 maps-related announcements (Gale 2013). Through these efforts, Google was able to gradually—but not entirely—wean itself off dependence on other maps data providers—TeleAtlas, Navteq, and Telcontar/deCarta (see Blumenthal 2009; Gale 2013). As a result, throughout this period, "Google Maps rose to become the de facto mapping platform for millions around the world. Fighting off direct challenges from rival technology companies such as Yahoo [and] Microsoft [. . .] (as well as from several regulators and government agencies)" (Gekker et al. 2018, 3).

This rise presented a number of challenges for other tech incumbents and for new entrants in that Google now held most of the maps-related cards, and its hand was only getting stronger. We see evidence of this strengthened position played out around its pricing structures for maps-related API access. For the first few years, Google Maps API access was free. In 2011, however, Google introduced fees for developer access to its maps. According to this newly implemented structure, sites would receive 25,000 free static map views per day, which is around 750,000 page views per month (DuVander 2011). Developers who used more than the free limit would begin paying $4 per 1,000 page views (DuVander 2011). In addition, sites were only permitted 2,500 free map views per day for "styled maps" (where users are allowed to change the colors and the visibility of certain map elements) (DuVander 2011). Developers who used more than their allocated allowance of 2,500 map views for styled maps would have to pay $4 per 1,000 page views up until 25,000; anything beyond this, the price then doubled (DuVander 2011). Marc Prioleau points out that, despite Google's claims to the contrary, the impact of these fees was potentially significant:

> I interviewed a guy at Google in front of a panel, and they just announced this [fee structure], and he was the head of Google Maps so I was going to grill him on this, and he said, "Well, you know, only 0.03% of the people using the Google Maps API will actually be affected by us charging," and that took me by surprise, and then I thought about it later and I thought, yeah, but [. . . that] 0.03% is like 30,000 [developers]. (Marc Prioleau, pers. comm., 2013)

With this one change, Prioleau notes, Google Maps "went from free to expensive" (pers. comm., 2013).

Then, in 2018, Google announced sweeping changes to Maps APIs, with "the 18 individual APIs Google Maps currently offers [. . .] consolidated into three broad segments—Maps, Routes, and Places" (Singh 2018). These API changes were accompanied by very significant fee structure increases, with Google "raising its prices by more than 1,400%" (Singh 2018).

Thus, for firms like Apple, who were likely to be among the afore-mentioned "0.03%," they were faced with the following decisions: work with Google (and the pricing structures they impose); seek alternatives (such as MapBox); work in partnership with other maps and maps data providers; or, forge their own path. Apple chose the last of these options.

APPLE

Apple's entry into mapping was marked by three distinct steps, or phases: the first was characterized by Apple's relatively quiet but steady series of strategic corporate acquisitions prior to 2012; the second, its much-publicized faltering launch of its "native" mapping app for iOS; and the third, the company's subsequent efforts to resolve these teething difficulties and move into its present expansion.

Establishment

Clear evidence of Apple's geolocation ambitions came with the purchase, in mid-2009, of Los Angeles–based mapping company Placebase, for an undisclosed sum (as well as reports of the poaching of talent from NextBus, a GPS predictive traffic data software start-up, and INRIX, a leading provider of street-traffic data [Markowitz 2012]). What made Placebase attractive to Apple, and also helped Placebase survive the arrival of Google Maps, was its ability to provide customers with map customizations, and to layer "commercial and other data sets (such as demographics and crime data)" onto maps using an application programming interface (API) called PushPin (Malik 2008).

Further purchases were completed the following year. In mid-2010, Apple acquired a Québec City, Canada, based mapping company, Poly9, for an undisclosed sum. Poly9 was noted for two features: its development of Mapspread, an online GIS tool that enabled collaborative map authoring; and, Poly9 Globe (formerly Free Earth), a competitor of sorts to Google Earth (as well as being the source behind the whimsical NORAD Santa Tracker site) (Dodd 2010; Forrest 2007; Nosowitz 2010). In addition to its own products, Poly9 also powered waypoint tracking for Garmin Communicator, which is a web browser add-on that transfers data between a Garmin GPS device and a computer (Dodd 2010). What is also notable about this purchase is that Poly9 obtained its geodata from Placebase's PushPin API, and one suggestion was that hiring a group of developers who were "intimately familiar the Placebase API" would be valuable to Apple should they produce "their own API for mapping that developers could use in iOS and/or web apps" (Dodd 2010). The purchase of Poly9 was also seen as a clear indication of Apple's interest in 3D maps (Gurman 2011).

In addition to Poly9, Apple also quietly acquired the Swedish 3D mapping company, C3 Technologies, for US$267 million in 2010 (although news of it did not break until 2011). C3 Technologies was a desirable target for

Apple, and a useful complement to the Poly9 purchase, in that the company had developed the capacity to build 3D maps as well as to "integrate 3D imaging into traditional 2D maps and other photographies" (Marshall 2011).

With each of these purchases, there was significant speculation in the US technology trade press regarding Apple's likely motives for buying them. While at least one commentator saw these moves as a clear signal of Apple's intention to "free itself from depending on Google for its maps" (Weintraub 2009), the prevailing sentiment seemed to be that any such move was far too ambitious ("I'm hesitant to think Apple has anything so difficult in mind as a Google Maps competitor," Nosowitz 2010; see also Prioleau 2012a, 2013b). More likely scenarios included the possibility that Apple might want to "launch their own turn-by-turn maps system, to remove one of Android's major advantages over the iPhone" (Nosowitz 2010), and that they might be aiming at adding in greater geotagging options for its own first-party apps (Arthur 2009; Nosowitz 2010).

However, by 2012, with the expiration of Apple's partnership with Google (Markowitz 2012), and with the announcement of a new partnership between Apple and the Dutch automotive navigation systems company TomTom (Preuschat 2012), it became increasingly apparent that Apple was indeed planning to launch its own maps.

Launch

The history of the development of Apple's mapping infrastructure tends to begin, in the minds of many, not with these early corporate deals and takeovers, but with the disastrous reception that followed the release of its maps app. This release was made as part of the launch, in September 2012, of the iOS6 operating system and the iPhone 5. Part of Apple's problems stemmed from the fact that the launch of its own maps application, which was quickly revealed to be riddled with errors and anomalies, was coupled with the company's decision to remove Google Maps for iOS from Apple devices, and, initially at least, to prevent it from being downloaded from the App Store. The end-user and third party developer backlash was swift and fierce. As Jason Farman explains,

> within hours of the launch of iOS6, the firestorm of complaints began to spread across the Internet. Within 48 hours, users had posted 245 screen captures from their iPhones to the Tumblr page, *The Amazing iOS6 Maps*, showing absurd glitches and misdirections from the Apple Maps app. These posts included a wide range of errors from geographical distortions in the Apple 3D perspective

to driving directions that took people across oceans and airport runways. Apple was also criticized for not including directions for users of public transportation. (Farman 2015a, 83)

There have been numerous efforts at interpreting what went awry for Apple in the lead-up to this launch and soon after. Many of the critiques that immediately followed the release of Apple Maps, Farman suggests, tended to accord with tech blogger Anil Dash's assessment that the only real beneficiary was Apple: "Apple made this maps change despite its shortcomings because they put their own priorities for corporate strategy ahead of user experience" (Dash, quoted in Farman 2015a, 83). For Farman, however, "frustration over the prioritization of corporate interests over user experience," while important, is not, in his estimation, the central issue here. Rather, "the key to the uproar over the shift in mapping interfaces [for end-users] was that it was a change in default maps happening on *mobile devices*" (Farman 2015a, 83). Farman believes that, in making this change, Apple lost sight of the fact that the mobile phone had become "the default interface for navigating everyday space" (84). This apparent misstep on Apple's part is unusual for a company that appeared to grasp very early on the significance of the intimate connection between mobile devices and individual identity construction and performance—something that was evident in its now iconic iPod marketing efforts (Bull 2007). It is also curious given that Apple has in the past tended to exert tight control over the release of new products and services, and has a strong reputation for developing clean and customer-friendly interaction design on its mobile devices.

Meanwhile, map industry experts interpreted the problems Apple faced at this time rather differently from the perspective just described. For these analysts, there were two key issues.

First, it was clear to some analysts that it *was* necessary for Apple to build its own mapping capabilities (Hern 2012). Marc Prioleau's (2012c, n.p.) argument is that "Apple could not continue to feed the Google marketing machine while continuing to compete tooth and nail for dominance of the mobile OS [operating system] market." This is an issue that will be returned to and expanded on later in this chapter.

Second, and at the same time as acknowledging they needed to compete with Google, it was also clear that Apple Maps was seriously flawed. Apple's issues were viewed by mapping industry analysts as stemming from a serious underestimation of the complexities associated with building meaningful geo services at scale and in such a compressed timeframe (compared, at least, with the length of time Google took to build its services). For instance, US maps consultant Mike Dobson (2012a, n.p.) lists five reasons

Apple would find it very difficult to build a high-quality mapping application. These included: challenges in providing high-quality map coverage outside of the United States; a lack of resources "to provide the majority of geospatial and POI [points of interest] data required for its application"; serious shortcomings with the companies Apple acquired to assemble its application (Dobson's view of these companies is that "they are, on the whole, rated 'C-grade' suppliers"); difficulties associated with amalgamating diverse business listings data from its three suppliers, Acxiom, Localeze, and Yelp reviews; and a larger challenge dogging all attempts at fusing business listings data—namely, the issue of "different approaches to localization, lack of postal address standards, lack of location address standards and general incompetence in rationalizing data sources."

In summary, Dobson (2012a) argues, "the issue plaguing Apple," both at the time of launch of Apple Maps and subsequent to this launch, was not "mathematics or algorithms," but "data quality." This, he argues, is how end-users ultimately measure the success or failure of a mapping application:

> Users look for familiar places they know on maps and use these as methods of orienting themselves, as well as for testing the goodness of maps. They compare maps with reality to determine their location. They query local businesses to provide local services. When these actions fail, the map has failed and this is the source of Apple's most significant problems. (Dobson 2012a, n.p.)

It is a perspective that accords with empirical research on map use, such as Didem Özkul's study of Londoners' "relationship between spatial behavior and locational information use" (2015b, 40–41). For many in Özkul's study, existing spatial knowledge formed a crucial precursor to, and touchstone for, people's subsequent use of mobile maps. However, it also became clear to Özkul that, increasingly, there was a shift from the use of landmarks (such as London Bridge) to the use of smartphone mapping applications for navigation and for the acquisition of spatial knowledge. When these services break down, mobile users, who increasingly come to rely upon them, are, as Farman (2015a) points out, quick to criticize.

To put Apple's mapping problems in more formal geolocational terms, priority areas for improvement included data completeness, logical consistency, and positional, temporal, and thematic accuracy (Dobson 2012b). Transit routing was considered another crucial omission that had to be addressed in later iterations (Prioleau 2012b) if Apple Maps were to improve.

Expansion

Despite these setbacks and challenges, Apple persisted with its mapping enterprises. In what David Kaplan (2015) calls Apple's current "expansion phase," the company has continued pursuing a similar strategy to that deployed pre-launch: further corporate acquisitions and quiet, behind the scenes development.

In 2013 alone, Apple bought five companies that have some relevance to the further development of its mapping infrastructure, with key additional purchases made in subsequent years.

The first of these, in March 2013, was the move for indoor location company WiFi Slam for US$20 million, which, it was suggested, would be part of efforts to allow iOS devices to "self-locate to within 2.5m accuracy in real-time" (Prioleau 2013a, n.p.). According to Prioleau, this level of accuracy represents the "Holy Grail" of indoor positioning due to its retail marketing possibilities. For Prioleau (2013a), however, the most likely use for such technology was not in developing indoor maps—a notoriously fraught and expensive undertaking—but in improving local search, and "in offering an indoor location API to app developers and combining it with a reminder service like Passport for iOS."

Then, 10 months after Grant Ritchie (CEO of Locationary) wrote an article describing five challenges Apple faces if it were to improve its mapping service (Ritchie 2012), his company was purchased by Apple—the second of its 2013 maps-related purchases. Described as "a sort of Wikipedia for local business listings," Locationary "uses crowdsourcing and a federated data exchange platform called Saturn to collect, merge and continuously verify" (Paczkowski 2013) large volumes of information on local business and points of interest to ensure that business listings data are positionally and temporally accurate (e.g., a restaurant is where it should be, and is still open for business). This had been a clear issue for Apple prior to this point, as a number of analysts had identified.

These purchases were followed by the acquisition of two smaller transport-focused start-ups: New York based HopStop, a transit-navigation service that draws from a variety of sources to show users the most effective way to travel by foot, public transportation, and car (Burrows and Frier 2013); and, Embark, a Silicon Valley firm that was financed by seed funds Y Combinator, SV Angel, and BMW iVentures, BMW Group's venture capital arm (Lessin 2013). Transit was another key area earmarked for urgent attention by mapping commentators at the time of the Apple Maps launch.

Apple also acquired BroadMap—although, more specifically, the acquisition actually involved staff and technology, not the actual company or its name (I. Fried 2013). These staff had "expertise that could assist Apple in improving the data it has for its Map app on iOS and OSX," including in "managing, sorting, and analyzing mapping data" (Gurman 2013). BroadMap's technology had also been integrated into products offered by MapQuest and Nokia HERE.

One of the challenges for industry analysts and academic researchers alike in examining Apple's corporate acquisitions is that the company's motives for making each purchase are not always immediately evident, and in many cases there are many possible applications for the technology that it gains access to through these purchases, which range across the various facets of the company's business. A good case in point are the two purchases Apple made toward the end of 2013, which, on the surface at least, appear to have little relevance to geolocation and mapping. Closer scrutiny, however, reveals a different picture. The first, reported in the same article that announced the BroadMap deal, was a little-discussed 2013 purchase by Apple of Catch, a note-sharing app and one-time competitor to Evernote. What is striking about this purchase in the present context is that the developers of Catch used to be responsible for an Android application called Compass, which was "location-based and infused location with notes" (Gurman 2013). Such an acquisition might suggest some form of integration between mapping and notes functionality on iOS. The second, in December 2013, and a surprising move to many, was Apple paying US$200 million for Topsy Labs Inc., a social media analytics firm known for its analysis of Twitter traffic. Speculation regarding its use to Apple focused on possible applications around iTunes, Apple's iAd products, or in refining Siri's search capabilities (Bercovici 2013). However, it was also noted that, at the time of its purchase, Topsy had "enhanced the ability to track tweets by the geographical location from which they were sent" (Wakabayashi and Macmillan 2013).

Following this flurry of corporate activity in 2013, Apple made further strategic purchases in successive years. In June 2014, for example, they acquired Spotsetter, "a social search engine built on top of a mapping interface" (Perez 2014). In truth, this was, once again, principally a technology and talent acquisition, including former Google Maps engineers Stephen Tse and Johnny Lee. Spotsetter developed a means of pulling data from users' content across social media platforms (Facebook, Twitter, Instagram, Foursquare), as well as venue content from 30 review sites (including Yelp, Zagat, the *New York Times*, Michelin, and TripAdvisor), to create personalized recommendations and reviews of nearby venues (Perez 2014)—valuable capabilities for enriching Apple Maps' content. In the same year, Apple also hired five people who had created (the now defunct)

Pin Drop, a mapping application that made it possible for users "to drop custom created tags and pins that could be shared with others to point out interesting landmarks and spots to visit" (Clover 2014). And, in mid-2017, Apple had posted more than 70 job listings related to mapping technology in one month alone (Kolodny 2017).

In April 2015, Apple purchased Coherent Navigation, the developers of High Integrity GPS (iGPS), a system that combines signals from mid- and low-earth orbit GPS satellites to "offer greater accuracy and precision, higher signal integrity, and greater jam resistance" (Slivka 2015). As Apple is not likely to be able to access iGPS technology—this was developed for the US military, and they are not a military contractor—the theory seems to be that Apple will draw on staff expertise within Coherent Navigation to develop accurate positioning systems in order to further enhance Apple's mapping capabilities (Oliver 2015). In addition, in September, Apple quietly acquired Mapsense (for around US$30 million), a San Francisco start-up "that builds tools for analyzing and visualizing location data" and that "lets users slice and dice graphical models of maps that hold huge sums of data" (Bergen and Chmielewski 2015).

"Apple acquisitions," one industry analyst writes, "are intriguing because they typically foretell future products" (Gurman 2013)—not to mention updates to existing products and services. This has been the case with Apple's map-related acquisitions, where, for instance, the tools and knowledge that came with Hopstop and Embark have been integrated into Apple's Transit feature that was unveiled with the release of iOS9 (Reisinger and Tibken 2015). There are also suggestions that the mapping technologies Apple is developing may well find their way into other products, such as the Apple Watch, possibly in the form of a dedicated GPS chip or other geo functionality linked to users' locations.

Not all Apple's work on improving maps during this period revolved around acquisitions, with the company busy on a range of fronts. To summarize the most significant of these efforts: Apple formed partnerships with TripAdvisor and Bookings.com (which supplement its existing deal with Yelp) to strengthen its recommendations, ratings, and venue data; renewed its license with TomTom; expanded a mapping-focused R&D+I center in Lund, Sweden (Wauters 2015); and launched its own fleet of cars with roof-mounted cameras (Buczkowski 2015; Gurman 2015).

In interpreting the implications of the last in the preceding list of developments, Gurman (2015, n.p.) writes that "for Apple, this means more control over the core user experience, and, consequently, less of a need to rely on partners for data to be updated," including, in the longer term, less reliance on TomTom for core map data, and on Yelp for images of storefronts and other venues.

Apple's efforts in improving its Maps did appear to be paying off, with, for instance, favorable reviews of Apple's iOS MapKit for developers (Grothaus 2013), and claims that Apple's 3D mapping products for OSX are now superior to Google's offerings, although the business use-case for the second of these products remains unclear (Braue 2013).

Despite the improvements Apple has made to its Maps services, many challenges remain. Arguably the most pressing of these concerns data quality. Concerns over the quality of Apple's base map data were raised as early as 2012, prior to the launch of Apple Maps, by mapping analyst Mike Dobson. Writing on TeleAtlas and its parent company TomTom, with TomTom being the "data supplier of most importance to Apple," Dobson makes the following observations:

> It is my sense that Tele Atlas has not prospered under TomTom ownership. TomTom's fortunes declined as the market for PNDs [portable navigation devices] unexpectedly, at least to TomTom, dropped shortly after the acquisition. Besides limiting the company's expenditures on improving the quality and coverage of TomTom's data, the drop in the amount of PND's sold decreased the update data available to Tele Atlas for map compilation purposes from TomTom's excellent Map Share product. Put another way, this is the company Google dumped because it was unhappy with the quality of the data delivered. (Dobson 2012a, n.p.)

These issues with (decreasing) data quality, Dobson explains, mean a range of supplementary agreements with other providers are necessary:

> In coverage areas where TomTom does not have the appropriate data, it appears that Apple will turn to other suppliers such as DMTI, a company that does provide relatively high-quality data for Canada, or Map Data Sciences, a company providing quality data for Australia and New Zealand. Unfortunately, other map data suppliers involved, in my opinion, do not meet these same standards and I would expect Apple's map data for much of the rest of the world to be lacking in detail, coverage and currentness. (Dobson 2012a, n.p.)

Dobson (2012a) goes on to suggest that Apple is likely planning to use the Google-owned Waze and perhaps OSM (OpenStreetMap) in situations where TomTom does not have sufficient data.

The "Acknowledgments" pages within Apple Maps for iOS8 give another indication of just how extensive Apple's licensing arrangements are with other companies in order to enrich their maps data. The difficulty for Apple is that, to maintain high levels of data quality, as mapping consultant Gary Gale (2015, n.p.) points out, "geospatial and mapping platforms need to be fed and fed regularly with data to keep them current." The unheralded arrival of its own street-mapping cars suggests at least one path that Apple plans to take to address this issue in the longer term. Another is the introduction in 2014 of "Mapping Connect," a "free, self-serve portal that allows businesses to add or amend their listings and related place-based content" (Gurman 2015).

Within only three years of its launch, Apple Maps, Marc Prioleau (quoted in Kaplan 2015, n.p.) suggests, is "much improved," and is no longer "a punchline to a joke." Nevertheless, data quality remains a live issue for Apple.

Apple and the Monetization of Maps

In wrapping up this examination of Apple Maps, I wish to address two intertwined questions that arise from the preceding material. First, why is Apple so committed to the development of its own mapping infrastructure given the obvious and significant challenges that it has endured and which remain before it? Second, how does Apple intend to monetize, to extract commercial value from, its Maps?

In entering the field of mobile mapping, Apple is clearly going head-to-head with the biggest of the mapping incumbents in Google. However, in developing this infrastructure, the underlying business motivations of these two big tech companies are somewhat different.

Google's primary economic interest is in driving search advertising—indeed, "Google's business *is* advertising" (Pon, Seppälä, and Kenney 2014, 983; emphasis added)—especially mobile-oriented, local search advertising. In order to achieve this aim, Google extracts data relevant to local advertising efforts from a variety of sources (Barreneche 2012b). This includes map and other place-related API data (Barreneche 2012b). It also includes information obtained via additional methods, such as geodemographic profiling that utilizes "the data-mining of records of location trails to produce the socio-spatial patterns that make up the segmentations that enable making inferences about users' identity and behaviour" (Barreneche 2012a, 339). And it includes information that is drawn from "a multitude of passive location logging technologies in order to amass a large and detailed geocoded data pool that can then be mined for commercial use" (Barreneche and Wilken 2015, 507).

Apple, in Marc Prioleau's view, "is still primarily a device and service business, not an ad platform, so their motivations [in developing mapping services] will be different from Google's" (quoted in Gurman 2015). This is not to say, however, that it is uninterested in developing ad service capabilities. Illustrative of this interest is the key (non-maps-related) acquisition that Apple made in early 2010: the purchase of US-based mobile advertising firm Quattro Wireless for US$275 million. This purchase was viewed at the time as Apple's response to Google's purchase the previous year of Silicon Valley–based mobile advertising firm AdMob Inc. (which Apple had also made a bid to acquire), for the eyebrow-raising sum at the time of US$750 million (Swisher 2010). A key motivation for the Quattro Wireless purchase, it's been argued, was Apple's desire to make money from its iPhone and (then yet to be released) iPad app platforms, which was a particularly important issue for Apple given that free apps heavily outweighed (and continue to outweigh) paid apps in its App Store downloads. As one commentator explained,

> Apple takes 30% of any revenues that third-party developers make from their paid apps, but it currently makes no money from free apps, which comprise the majority of those that are downloaded. (Iwatani Kane 2010, n.p.)

It has been suggested that the "value of app stores for platform owners is not principally the income they generate, but rather the opportunity they provide to cultivate and manage an ecosystem"—they are "ecosystem control points, rather than revenue sources" (Goldsmith 2014, 175). Nevertheless, because "many free apps run mobile ads, [. . .] a mobile-advertising service could give Apple a revenue stream from some of those apps" (Iwatani Kane 2010, n.p.).

Another likely motivation for Apple's investment in its own mapping infrastructure concerns its desire to minimize the extent to which it feeds Google's advertising profits. Looking at the third-quarter financial results for 2012, the reporting period when Apple Maps was launched, it is clear that Apple was clearly outpacing Google with regard to overall revenue (US$35 billion for Apple, compared with US$14 billion for Google), with Apple generating the lion's share of its revenue from device sales. For Google, the main source of its revenue was advertising, including mobile advertising (Google 2012; "Apple Reports Third Quarter Results" 2012), and map use is a vital component of successful (especially local) mobile advertising and associated revenue generation. Here the issue of "platform control," or "operating system-based power," becomes crucial.

In 2012, when Apple Maps was launched, Google's Android operating system (OS) and Apple's iOS were "deployed on 85% of smartphones sold" (Goldsmith 2014, 173). While Apple's aim with its OS was to try to "cultivate and manage an ecosystem" (175), Google's approach was somewhat different:

> The end goal for Google may be [. . .] about ensuring that as many devices as possible are able to run the most recent version of Google's services, such as Google Maps. (Pon, Seppälä and Kenney 2014, 987)

Who has home-screen advantage in terms of mapping applications becomes a key issue, and it makes sense that Apple would wish to minimize its reliance on a major player and competitor like Google for a vital service like geolocation. While it might hearten Apple executives to learn that "Google Maps on iOS has 100m or so monthly active users, out of perhaps 500m iPhones in use" (B. Evans 2015), the fact remains that, based on 2014 figures, Apple has almost 56 million fewer users of its maps app than Google does (McDermott 2014).

Maps app battles aside, another potential threat to Apple's nascent geolocation ambitions looms on the horizon in the form of Facebook (see Chapter 3). With the purchases of Instagram and WhatsApp, Facebook, as we shall see in the next chapter, has made two very shrewd acquisitions. Not only are both key image-sharing and instant messaging services, with the capability for transmission of a wide range of content (Lobato and Thomas 2014, 119), but much of this content can be and is geocoded. With their purchase comes significantly strengthened geodata capture capabilities for Facebook—none of which would have escaped the attention of Apple executives. Thus, it would appear that there is some substance to suggestions that part of the motivation for Apple in building its mapping capabilities was to add greater geolocation options to its own first-party apps (Arthur 2009; Nosowitz 2010) and to open up further revenue options (van Dijck 2011, 343) by enabling geolocation functionality. What these options are likely to be, however, are far from being fully resolved.

CONCLUSION

In this chapter I have examined one of the newer arrivals on the mobile mapping scene: Apple. In looking at Apple, I have traced the early development, troubled launch, and more recent consolidation of Apple Maps. Lacking the necessary initial resources to develop quality mapping infrastructure, Apple set about building this capability through successive waves of technology, "talent," and corporate acquisitions. Despite an infamously rocky start, these investments, accordingly to recent maps industry analyses, appear to be paying dividends for Apple. While Apple has always remained evasive about its plans for its maps, there are clear motivations for why it might wish to build this capacity. As a company, Apple is commonly perceived as a manufacturer of slickly designed devices that form part of an "Apple 'ecosystem' driven by hardware" (Lobato and Thomas 2014, 120). However, given the significant revenue generated by its own devices, Apple is naturally eager "to make sure it can make as much money from its products as possible" (Frommer 2010), and, where able, to maintain control of its core services and key monetization opportunities through "walled garden" business models and "vertical integration." With the growing significance to its business of mobile mapping, and to a lesser extent mobile marketing, it makes sense for Apple to be cautious about ceding control of these two core functions to key rival Google, and for it to try and maximize its own monetization opportunities.

All of this, however, is shaping up to be a longer (rather than short- to medium-) term strategy. With maps, there are no quick fixes. In an interview with Matthew Panzarino of *TechCrunch*, Apple SVP Eddy Cue, who now oversees Apple Maps, revealed that, rather than rely on "a patchwork of data from partners [. . .] and other geo brokers," Apple has been "rebuilding the maps part of Maps" itself (Panzarino 2018, n.p.). It has been doing this by using "first-party data gathered by iPhones with a privacy-first methodology and its own fleet of cars packed with sensors and cameras" (Panzarino 2018, n.p.). For Panzarino (2018), "this is nothing less than a full-reset of Maps and has been four years in the making." Apple's new map is being rolled out gradually, beginning with the San Francisco Bay area. Apple's ultimate goal is to own the "entire map stack," as Google does, which means that it owns all the location and base map data it collects (O'Beirne 2017; see also Buczkowski 2018). All the while, Google's position continues to strengthen, with suggestions that Google's Maps-related Promoted Pin ads "have the potential to add [US]$1.5 billion to Google's revenues" in 2017 alone (Buczkowski 2017b, n.p.). In addition, analyst Justin O'Beirne (2017) draws from his own extraordinarily detailed side-by-side comparison of Google Maps and Apple Maps to conclude that Google's mapping advances are such that it has created a "moat of time" around its maps offerings, which protects Google from its competitors, giving it a "6+ year lead over Apple in data collection" at least. With Apple's "full-reset," this moat is only going to get wider.

In the next chapter, I continue this exploration of the political economic dimensions of location-based services, shifting focus from maps to mobile social networking firms, and examination of the integration of location into their services, and associated business and revenue models.

CHAPTER 3
Location Integration and Data Markets

Foursquare: The point isn't to become profitable right now, the point is to grow as quickly as possible, to really push the boundaries of what you can do with location based services on mobile devices.
> —Dennis Crowley (quoted in 140Talks 2011, n.p.)

Facebook: A location service only gets interesting when you get to a certain scale.
> —Josh Williams (quoted in Constine 2012c, n.p.)

INTRODUCTION

In an oft-quoted adage, generally attributed to Silicon Valley entrepreneur Steve Blank (2010), a start-up is defined as "an organization formed to search for a repeatable and scalable business model." This is an apt characterization of many of the key locative and social media firms that were launched—post the "dot.com" crash—in the first decade of the 2000s, such as Foursquare and Facebook, among many others. These firms were born into what writer Michael Lewis (2001) would term a post-Netscape era of fluid business models, where tech start-ups are backed by venture capital and other forms of investment, while the developers and investors set about figuring out how best to monetize their assets (see also Lacy 2009). Common and recurrent questions asked of these technology start-up companies include: What are the markets they are servicing? What are their underlying business models? How do they intend to raise revenue and seek to generate profit? These are shared concerns for a range of stakeholders: investors and other business-oriented analysts trying to gauge the likely success of these new enterprises and their suitability as investment and/or acquisition targets; media and communication scholars interested in changing patterns of media economics, media industries, the significance of the new types of mediated practices and services these new companies support, and the production and consumption of media content; and, policymakers charged with formulating or applying the regulatory frameworks under which these companies will operate. Asking these questions is also important if we are to move beyond the techno-boosterism that all too often accompanies the emergence of new types of media and technology business—and the associated sweeping claims of how they will "disrupt" established industries and revolutionize established practices of everyday life—to arrive at a more considered analysis and evaluation of the development and significance of these firms.

In this chapter, I examine the still-evolving business and revenue models, as they relate to the geolocation data capture efforts, of two commercial businesses now central to the contemporary settlement of locative media. The analysis is comparative, detailing the efforts at geolocation capture and integration and revenue generation of a company that is clearly identified as offering location services (Foursquare), and of another that, until relatively recently, has not generally been thought of as location-based (Facebook). In developing this chapter, I draw extensively from, and further extend the arguments of, prior published work of mine on Foursquare (Wilken 2015, 2016; Wilken and Bayliss 2015) and Facebook (Wilken 2014a). My examination of these two firms is not intended to be exhaustive. Rather, it seeks to draw attention to a series of important moments and key developments as both firms set about building their location capacities, and adjusting their business and revenue models accordingly. These two case studies provide a rich composite picture of the ecologies of locational information. These cases also aim to develop a clearer understanding of how both firms accrue location *data* and, from this, how they extract location *value*—that is, how this information is shared, harvested, valued, reused, and commodified.

FOURSQUARE

As an established player in the field of location-based services, Foursquare presents a valuable case study for considering the logics and assumptions informing the development of mobile location services—in particular, the approaches and strategies that have been adopted, and the ongoing evolutions that have been necessary, in order to build a viable and sustainable business model.

Foursquare is a location-based mobile social networking and, more recently, search and recommendations service. It rose from the ashes of Dodgeball, the pioneering mobile service that New Yorkers Dennis Crowley and Alex Rainert created in 2000 and subsequently sold to Google in 2005. Crowley left Google when it became clear that Dodgeball was not going to receive further engineering resources from Google; Dodgeball was eventually shelved by Google in 2009, replaced by Google Latitude, and rolled into its Google Places API. Determined to continue developing the Dodgeball concept, in 2009 Crowley and Naveen Selvadurai founded Foursquare, with Rainert joining soon after. Crowley and Selvadurai chose to base their company in the technology start-up cluster emerging in New York City, the so-called Silicon Alley (Neff 2012), which would seem to be a natural fit for the urban, pedestrian-focused locative media service. This service grew to become a key player in the area of location-based mobile social networking and local search, with Foursquare reporting they had attracted in excess of 40 million users by 2013 ("About Foursquare" 2013), up from 10 million in 2011 (Gobry 2011), and over 4.5 billion total check-ins ("About Foursquare" 2013), up from 1 billion in 2011 (Shontell 2011). By 2018, the number of users of its apps is said to have surpassed 50 million per month, with 12 billion total check-ins ("About Us" 2018).

What set Foursquare apart from its competitors when it first launched, and was of particular appeal to its early adopter heavy users, was the emphasis it gave to its various gameplay elements, where each Foursquare user collected badges for venue check-ins, competed with friends over a check-in leader board, and with other users to become "Mayor" of venues. Each of these gameplay aspects is explained briefly in turn.

First, individual users could collect a variety of merit-style badges. Often whimsically titled, these were scaled to reward various progressive levels or stages of user engagement. So, for example, new users could achieve the "Newbie" badge before progressing, following heavy enough check-in use over a given time, to unlocking the "Super User" badge, or the "Swarm" badge when a check-in was received in close temporal proximity to those of over 50 other fellow Foursquare users. In late 2011, Foursquare also introduced scaled achievement levels within each badge (as well as a small suite of additional badges) so as to, in their words, reward venue exploration and help show individual user "expertise" ("Level Up" 2011). This meant, for instance, that an occasional café-goer might achieve level 2 of the "Fresh Brew" badge, whereas a café-frequenting coffee aficionado might achieve level 10 of the same badge. As far as Foursquare is concerned, the second user is of far greater interest in terms of the check-in information and recommendations data they contribute to the service's metrics.

Second, the app featured a dynamic table in the form of a constantly updating leader board, that mapped who, in a given user's social network, was achieving the most check-in points over a seven-day period. The aim was to encourage playful competition between members of a user's social network and, presumably, drive up the number of total venue check-ins. Some venues also offered discounts and other deals for check-ins.

Third, Foursquare encouraged users to compete with each other to become "Mayor." This was the honorary title given to an individual user who has checked in most frequently to the same venue over a 60-day period. Foursquare's gamification integration was so successful that it was rapidly replicated by other competing services. For instance, soon after Foursquare launched, Yelp introduced a "royalty" system of its own. Rather than become "Mayor" of a venue, Yelpers were competing to become Duke/Duchess (most check-ins to a venue), Baron/Baroness (most titles in a neighborhood), and King/Queen (most in a city) (Siegler 2010). Within the tech sector, numerous other companies have tried to follow Foursquare's (and Yelp's) lead, with Facebook, the Google-owned crowd-sourced traffic information service Waze, language-learning platform Duolingo, communication app Line, and numerous others, all incorporating game elements into their operations (Mishra 2014).

While Foursquare used these gameplay features to drive user growth, the company faced a much tougher time in the few years that immediately followed the launch of the Foursquare application. In early 2011, Foursquare Labs Inc. made a much-publicized strategic shift in direction that has taken it away from its prior emphasis as a location-based mobile social networking app driven by game dynamics. This was a decision generally regarded as a response to persistent questions the company faced by industry analysts questioning the long-term sustainability of its business (Isaac 2013). It was also a result of reported slowing in user growth, including in emerging markets outside of the United States (B. Evans 2013), such as Indonesia and Turkey.

Faced with these challenges, the company radically rethought its corporate strategy, specifically by redesigning the application itself and the services it offered end-users, and by further honing its still nascent business plan, including by building services to cater to business. The result was that Foursquare appeared to turn away from leader boards, badges, and points, to focus instead on local search, discovery, and recommendations. Foursquare Labs Inc. unbundled its operations. In reality, what Foursquare did was spin-off the gameplay/check-in aspects of its service into a new app called Swarm. Even here, as I have argued elsewhere, the gameplay elements were temporarily squeezed out, but then were reintroduced in later updates (Wilken 2016), with a subsequent revamp undertaken to accommodate "lifelogging" and the recording of one's personal locational traces (Crook 2017b). Meanwhile, the original Foursquare app was redesigned as a dedicated search and recommendation service, known as Foursquare City Guide. As Foursquare's then head of business development, Holger Luedorf, put it at the time, "we're positioning ourselves as the location layer of the Internet" (quoted in Panzarino 2014, n.p.).

The company's plans, as Crowley described them, were to "get most of [its] future sales from software that helps merchants track the behavior of potential customers"' (Crowley, quoted in E. Chang and MacMillan 2011, n.p.). Foursquare already collected some revenue through strategic partnerships with competitors and a variety of companies (E. Chang and MacMillan 2011; Van Grove 2011, 2013)—most significantly, a US$15 million deal with Microsoft that saw the software giant making "substantial" additional regular payments to Foursquare for access to its proprietary location data (Tate 2014).

Foursquare's merchant platforms, however, were quite different in that they encouraged businesses to pay for help in analyzing the data generated by users through Foursquare's service (E. Chang and MacMillan 2011). In 2012, Foursquare also launched "promoted updates," a service that allows its business clients to send advertising messages to users who are in the vicinity of a restaurant or other venue (Kelly 2012). A further merchant feature was the Foursquare for Business app (Foursquare Blog 2013b; Isaac 2013). Launched in early 2013, it allowed businesses to offer deals when users checked in, as well as enabling them to send messages to regulars (Isaac 2012). Additionally, by October 2013, the company had also opened up Foursquare Ads to small businesses around the world (Foursquare Blog 2013a), and, a few months later, had partnered with ad tech company Turn to deliver ads to its users on desktop computers, tablets, and mobiles (Delo 2014).

Key to the success of these corporate-focused initiatives was a major redesign of the actual Foursquare app around the "Explore" feature. In essence, Explore functioned as a recommendations and ratings system that utilized a series of metrics drawn from each user and his or her social network history, including tips, likes, dislikes, popularity, local expertise, and so on (Kerr 2012). This information was then targeted to that user, in Foursquare's words, in the form of "recommendations for places that you would probably like to visit based on your profile and check-in history" (Goldman 2012). In a second development for its end-users, in late May 2013, Foursquare added what it called "super-specific search" to Explore (Welch 2013). This applied a range of filters to search results that combined common queries (such as price, opening hours, and so on), with additional information drawn from check-ins and user data. By September 2013, restaurant menu search capabilities had also been added (Sterling 2013b).

Foursquare's ambitions for Explore extended beyond the compilation of location information of this sort, to also combine mobile, social, and location-based interactions with past and present user data to generate real-time and even predictive recommendations. Foursquare's chief data scientist, Blake Shaw, explained how the company's engineers work with and combine two interconnected data sets in order to make predictive recommendations: social data (Foursquare's "social graph"), and location-related data (its "places graph"). Both "graphs" are composed of "nodes" (things—people and places) and "edges" (the connections between these things) (Shaw 2012). Within its social graph, nodes are Foursquare users, and edges are the connections linking these users to each other, including as a result of friendships, follows, "dones" (tips posted by one user that other users then do), comments left by users, and the "co-location of people [who are on Foursquare] in the same physical space" (Shaw 2012). In the places graph, the nodes are places that are registered in its points of interest database, and the edges are composed of a variety of different things: flow ("how often people move from one place to another"); co-visitation ("how many people have been to the same place before"); categories (the sorting of venues based on similarities between them); and menus, tips, and shouts (which are described as data "which connects places because they share the same characteristics") (Shaw 2012, n.p.).

Foursquare's ambition has been to combine these data sets and develop from them responses to queries generated via the Explore feature. These responses are created in order to produce for users "realtime recommendations from signals [that combine] location, time of day, check-in history, friends' preferences, and venue similarities" (Shaw 2012, n.p.). The larger ambition of Foursquare's engineers is to better understand the points of intersection between these two graphs. As Shaw (2012) asks, "What are the underlying properties and dynamics of these networks? How can we predict new connections? How do we measure influence? Can we infer real-world social networks?"

Foursquare's ability to answer these questions would depend in large measure on the ongoing population of its places database, and the company took a number of steps to try to facilitate continued and expanding supply of geolocational information flowing into its place database.

First, in targeting consumer end-users, Foursquare introduced at least three subtle adjustments to its mobile interface that aimed to encourage "frictionless location sharing." First, there was the quiet removal of the ability for users of its iOS version to check in privately; in doing this, Foursquare ensured, in the words of one industry commentator, that "check-in data is accessible to users of the product, its API partners, and any possible suitors for acquisition" (Panzarino 2013, n.p.). Second, for its iPhone app, a series of unobtrusive venue-related user feedback questions was introduced (Foursquare Filling 2013). These questions pop up after one checks in to a location, and prompt users to respond to such questions as, "is it quiet here?," "would you grab a quick bite to eat at this venue?," "does it have Wi-Fi?," and so on—feedback that enabled the company to further populate its places database with crowd-sourced "rich" venue data (see Figure 3.1). Third, Foursquare also sought to position Explore as a "passive venue search system" (Shaw et al. 2012, n.p.), and the company encouraged iOS users to activate push notifications on their phones for nearby venue recommendations and always-on location in order to move away from the

Figure 3.1. A pop-up reminder from Swarm app of the "automatically saved traces" it collects and the recommendations it offers based on these.
Screenshot: author.

prior reliance on users being the ones to initiate interaction with the service. These tweaks to the interfaces of its flagship mobile application to encourage greater venue and location sharing were part of Foursquare's ambition to build a *predictive* mobile search and recommendation service.

Second, Foursquare also continued its support for engaged super users by transitioning the 4sweep service to a new server in 2018, as well as adding this tool to software development service, GitHub; 4sweep is a Foursquare API for finding and reporting "duplicate, closed, miscategorized, private, and inappropriate venues, photos, and tips" (4sweep 2018, n.p.). While 4sweep is rarely discussed in the trade press coverage of Foursquare's operations, it is significant in that it provides an outlet for crowd-sourced population of up-to-date venue data within its places databases.

Third, Foursquare has continued to enrich its places databases by maintaining an "open" API. This has been strategically important to Foursquare since its inception (Wilken and Bayliss 2015). By offering third-party developers a way to access parts of a company's data, "APIs have also become a useful way for these companies to extend their reach and growth across the Web" (Bucher 2013a). This is particularly important for Foursquare, as API usage tends to generate more traffic than end use does (Bucher 2013a; Sheehan 2013).

As a result of all of these developments, Foursquare Labs Inc. has consolidated both its position in the location-based services landscape and its identity as a firm, and its business and revenue models are starting to come into sharper focus (Gobry 2012).

For consumer end-users, Foursquare Labs Inc. can now be understood in two ways. On the one hand, it continues to provide a gamified location-sensitive mobile social networking and check-in (or "lifelogging") service (Swarm app). And, on the other hand, it also provides a search and recommendation service through its flagship Foursquare City Guide app—one that is now in direct competition with Yelp and Zomato. In 2016, Foursquare Labs Inc. sought to further strengthen this aspect of its consumer-facing operations by launching Marsbot for iOS. Marsbot, an app that represented a return of sorts to the early days of Crowley's pioneering SMS service Dodgeball, is a recommendations bot that sends users "contextually aware, proactive recommendations" for venues by text message (Chacko 2016; Levy 2016; Newton 2016a).

For the tech industry, Foursquare Labs Inc. has evolved to become an important SaaS (software as a service) firm, providing location intelligence for other tech companies and developers. This service takes the form of two offerings: its Pilgrim SDK (software design kit), proprietary software that is said to be able to sense when phones stop at or enter a venue (Crook

2017a; Flynn 2017; Johnson 2018; Rosenblatt 2017; Yeung 2017); and its Places API, which provides other platforms and developers with access to its points-of-interest database. In the past, Places has provided location data for Yahoo!'s Flickr, Evernote, Kakao's Path, Twitter's now discontinued Vine, and, very early on, for Google's Waze and Facebook's Instagram ("Announcing Foursquare & Button" 2015; Calore 2013; Carr 2014; Dash 2013; "Introducing" 2013). Now, Places is said to "power location data for Apple, Uber, Twitter, Microsoft" ("About Us" 2018, n.p.), as well as Mapbox (Gundersen 2018), Tinder (Kaplan 2018), Snapchat, Instacart, and Lonely Planet (Fingas 2018), a host of car companies (Foursquare 2018), and "100,000 other developers" ("About Us" 2018, n.p.).

And, for wider industry, Foursquare is increasingly understood as a location analytics and ad platform. With respect to its ad offerings, Foursquare provides two services: Pinpoint, a location-based targeted advertising tool (Lopez 2015); and, Attribution, a location analytics tool for measuring foot traffic and the impact of online and traditional advertising (Sterling 2016) so as to determine LROI, or location return on investment (Sterling 2015).

To close this discussion of Foursquare, I and others have previously suggested that Foursquare's decisions to unbundle its service, reconfigure its business and revenue models, and partner with other tech industry players in order to enrich its places databases have constituted risky strategies for the company to take. However, increasingly, these risks appear to have been worth taking (Crook 2018). The net result of efforts over the past decade is that Foursquare has now built up a "huge database of location names and addresses," and this points-of-interest database "has become an essential part of the location-based application ecosystem" (Frith 2015b, 105), one that is intrinsic to its "existing forcefield of institutional structures and functions" (Schiller 1999, 35). Indeed, as Barouch (2013) has argued, such is the richness and importance of Foursquare's places database, that, now, "any destabilization of Foursquare or its developer tools would fundamentally affect the stability of the mobile web" (Barouch 2013, n.p.). Foursquare, it would seem, is a step closer to fulfilling cofounder Dennis Crowley's desire for it to become *the* "location layer of the internet" (J. Edwards 2013, n.p.).

The following section considers how Facebook set about building geolocation data capture capacity, and integrating it into its services. My aim here is not to develop an exhaustive account of all of Facebook's work on geolocation. Rather, my focus is on accounting for a number of key developments between 2010 and 2014, a period when Facebook focused much of its attention on building the capacity to capture geolocation information and integrate it into its services.

FACEBOOK

By Mark Zuckerberg's own admission, Facebook was a relative latecomer to location and mobile location. By the time it launched its first significant location offerings in 2010, other major players, such as Google (see Chapter 2) and Twitter (Wilken 2014c), had already integrated location into their operations (Bilton 2010). In addition to Google and Twitter, a suite of specialist location-based mobile social networking start-ups had also emerged, including Loopt (founded in 2005), Whrrl, Brightkite, and Gowalla (all founded in 2007), and, as discussed earlier, Foursquare (founded in 2009 from the ashes of Dodgeball).

Despite playing catch-up, Facebook played a cautious hand. Rather than rushing to add more explicit locational elements to its social networking service, Facebook focused on laying careful foundations on which to build meaningful location functionality. An announcement about its much-rumored location features was expected at its f8 developer conference in April 2010 (O'Dell 2010). However, clearly concerned about the possible privacy implications of aggregating users' location data, this announcement was delayed in order for the company to "hammer out" a new privacy policy (O'Dell 2010). An earlier update to the company's privacy policy, released in November 2009, paved the way for geolocation integration by inserting the statement: "When you share your location with others or add a location to something you post, we treat that like any other content you post" (quoted in Bilton 2010, n.p.). Care was also given to achieving the seamless integration of location within the Facebook platform in a way that would accord with its business plans while not disrupting end-user engagement.

Facebook Places

The result, Facebook Places, was eventually unveiled in August 2010. There were two aspects to this new location feature. The first, a mobile-only service, allowed Facebook users to check in via smartphone to specific locations and to share each check-in with friends. The decision to launch a mobile-only feature was likely driven by acknowledgment that most competing location-based check-in services were mobile driven, and in recognition of the fact that, as of 2010, Facebook had "200 million people around the world [who were] actively using Facebook from a phone," a number that had tripled from the previous year and was only likely to continue growing (Tseng 2010). With this in mind, Facebook tried to sell Places to its users via its blog with the pitch that "life happens in real time, and so should sharing," including the sharing of specific locations (Tseng 2010).

Further Places refinements followed. A "starred friends list" was added, which was a way of tagging those friends a user frequently checked in with. A Places Editor app was also tested that enabled users to correct location or venue information and categorize this information (Constine 2011a). Facebook also made the Places service more attractive for both its end- and business-users by introducing Deals (Tseng 2010)—that is, offers that users could share with friends nearby (Fougner 2011). Deals was initially launched in the United States, and subsequently expanded the following year to include Canada, France, Germany, Italy, Spain, and the United Kingdom (Fougner 2011). Deals formed a key addition to Facebook's location offerings: while Facebook claimed in 2009 that any location service it might develop would not compete with Foursquare, Loopt, or Gowalla, with Deals it was clearly beginning to stake out similar turf as two other competitors: Yelp and Groupon.

The second aspect to the Places feature was Facebook's decision to open up its place editing API and geocoding service to a limited number of third-party developers. Significantly, this permitted the ability for a select list of "intrepid developers (including Foursquare, Gowalla and Loopt) to develop interesting location-based services on top of Facebook" (Lardinois 2010, n.p.). It was a particularly canny move on Facebook's part. As a company without first mover advantage in the area of geo services, opening up its API enabled it to gain access to, and aggregate location data generated through, other applications. Thus, its Places database grew exponentially.

Even more powerful was the addition of single sign-on to its mobile app (Tseng 2010). Otherwise known as Facebook Connect, this service gives smaller sites the option of allowing their users to sign in via their Facebook account (Bilton 2010). The attraction for users was that it simplified authentication processes, and enabled them to "'connect' their Facebook identity, friends, and [apply their Facebook] privacy [settings] to any site" (Morin 2008); interactions that occur on these other sites would then also appear on users' Facebook pages. The attraction for businesses was access to "the precious user data connected to the platform," including, with certain permissions, users' location information (Hijleh 2012)—information that marketers tend to view as highly prized. There were also obvious benefits of single sign-on for Facebook. Given that the announcement of single sign-on for mobile made explicit mention of Loopt, Yelp, and Groupon, among other apps (Tseng 2010), one clear benefit would appear to have been granting Facebook access to a large pool of geocoded data, much greater than Facebook's users would generate via the Facebook app alone. Reflecting on one description of single sign-on as "like a virtual passport," Nancy Baym (2011, n.p.) asked whether "we really want to think of Facebook as a nation," and questioned the implications of what it meant for "Facebook citizenship to become a requirement for accessing other domains." Facebook's appeal to the concept of the nation in selling single sign-on was telling insofar as this feature of its interface evoked another concept from political theory: that of the "sphere of influence," which, in its loosest sense, is used "to denote any territory in which a foreign power sought to exert exclusive influence without annexation" (W. Moore ca. 1963, 165). When applied to Facebook's business dealings, this was precisely what single sign-on was designed to achieve: a means of exerting influence over how web-based information is accessed in order to gain privileged access to the data—including geocoded data—that restricted access yields.

The "Afterlives" of Facebook Places

By August 2011, only one year after its launch, Facebook Places was discontinued. Jessi Hempel makes reference to Facebook's "three steps forward one step back launch approach," where the company launches a product in order to test the waters, then pulls back before rolling it out again more slowly (cited in *Mark Zuckerberg: Inside Facebook* 2013, n.p.). As this remark suggests, the discontinuation of Places by no means signaled a diminished interest in location and geodata on Facebook's part. On the contrary, at the same time that it was "killing Facebook Places," the company was "adding a lot more location features" (Protalinski 2011a) and revamping its privacy settings in order to accommodate them (Constine 2011b). Facebook also enabled location check-ins for its desktop users (rather than limiting it to mobile-only functionality), and permitted opt-in location-tagging of all Facebook content (status updates, photos, Wall posts, and so on [Protalinski 2011a]). The preceding were all key steps in Facebook's vision of location as metadata, where "location isn't just a node in the graph, but information that could be part of all content" (Tseng, quoted in Constine 2010, n.p.), a "layer" (Constine 2012e) sitting over the top of everything.

It is possible to interpret these developments in a number of ways. The move by Facebook away from a mobile-only location service could well have been driven by awareness that, while mobile internet use was growing, it still represented a smaller proportion of wider internet use. For instance, in 2010, of the overall proportion of the US population who accessed the internet (79%), 29% did so via a cell phone (Zickuhr 2013). Second, it is also possible that take-up was slow, and that Facebook's mobile users were reluctant to geotag content. By expanding location-tagging capabilities to desktop users, Facebook would be able to target a much larger proportion of its overall user base and, in the process, habituate users to the practice of geotagging content (Protalinski 2011a). Third, by no longer restricting location-tagging to the present, and widening this capability to include the past (such as geotagging old photos) and the future (such as sharing tips with friends regarding future events), Facebook had the potential to expand significantly the pool of user-generated geocoded data.

In this interim period, Facebook engineers also worked on a new way of presenting check-in data: the "timeline map." This required back-end work, building a "global places directory," along with data-fetching capabilities and aggregation algorithms, so as to achieve accurate location pins on a map, as well as systems to enable users to retroactively geotag their content in order for it to appear on the timeline (Mangla 2012).

In order to realize their ambition of the "timeline map" constituting "a single source for people to display the places they've visited," the Facebook engineers also released a series of location-related APIs that opened up access to third-party developers (beyond the select few of Loopt, Foursquare, Gowalla, and so on). These included APIs for "read, write, and search." In the case of the first of these, the *read* APIs, with permission, "will allow any application that a person is interacting with to access the places that that person and their friends have visited" (Mangla 2012). Moreover, as Constine (2012d, n.p.) explained, the read API "lets developers pull the coordinates of your friends based on their posts from Facebook or any location [. . .] as long [as] they're cross-published to Facebook and you're authorized to see them there, you could view Foursquare posts on Highlight, or Banjo posts on Glassmap." In the second case, *write* APIs "allow applications that have obtained user permission to post content and location tags directly onto [a user's timeline] map" (Mangla 2012, n.p.). While in the third case, *search* APIs were upgraded "to allow applications to access universal search requests so they don't need to build their own location search capabilities" (Mangla 2012, n.p.). For Constine (2012d, 2012e), these developments reposition Facebook as a "hub for location data," a "backbone" that can carry extensive social interaction and wide geofunctionality, and represent a further significant scaling of its location ambitions and a further extension of Facebook's sphere of influence in the field of location-based services.

FACEBOOK NEARBY

The next major step in Facebook's engagement with location came with the launch of Nearby for iPhone and Android in December 2012. Why Nearby is important, and how it differed from Places, is revealed by understanding three key, strategic corporate acquisitions Facebook made around that time.

The first of these was the purchase of Texas-based Gowalla in December 2011 for an undisclosed sum. Gowalla was an early location-based mobile social networking service that was understood to have lost significant ground to its New York–based rival Foursquare (Protalinski 2011b), making it a prime candidate for acquisition. Facebook's interest in acquiring Gowalla was not the service itself, which it closed within months. Rather, the key motivation for buying it was gaining access to the expertise of its staff, many of whom, including cofounders Josh Williams and Scott Raymond, relocated to Palo Alto, California, to work on further building Facebook's location services (Protalinski 2011b).

The second key acquisition was Instagram. In early April 2012, two months after filing its own initial public offering (IPO) paperwork but still not yet a publicly listed company, Facebook purchased the popular mobile photo-sharing site, Instagram, for US$1 billion (Constine and Cutler 2012). A week before this purchase, Instagram itself had closed a financing round worth around US$50 million (Tsotsis 2012a). The price Facebook paid for Instagram was considered high, even by Silicon Valley standards. According to one rather blunt industry assessment, the reason Facebook was prepared to shell out so much for the company was clear: "Facebook was scared shitless and knew that for the first time in its life it arguably had a competitor that could not only eat its lunch, but also destroy its future prospects" (Malik 2012). This, it was suggested, was due to the fact that Instagram not only had a passionate user base ("People like Facebook. People use Facebook. People love Instagram"), but, more crucially, because Facebook was "essentially about photos, and Instagram had found and attacked Facebook's achilles heel—mobile photo sharing" (Malik 2012, n.p.). Instagram, in short, had "cracked the code where Facebook itself failed: viral growth on mobile" (Malik 2012).

The third key acquisition was Facebook's purchase in May 2012 of Glancee for an undisclosed sum (Tsotsis 2012b). Glancee was one of a number of "second-generation" location start-ups known as "ambient social location" or "social search" applications (Lee 2013, 2728). Glancee tracks a user's location in the background, links to Facebook and Twitter accounts, shows "people who are using the app and their shared social graph interests and

Facebook picture" (P. Burns 2012) and includes a "radar" function to reveal their physical proximity (Lee 2013, 27).

Facebook's purchase of Glancee was also principally a "talent acquisition." The application was shut down, and Glancee's three cofounders, Andrea Vaccari, Alberto Tretti, and Gabriel Grise, all joined Facebook. Labeled "a nice-guy ambient social location app for normal people" (Eldon 2012), this acquisition was viewed at the time as a good fit between Glancee's "ideas and founders" and Facebook's "mainstream user base" (Tsotsis 2012b). As a "talent acquisition," Facebook would have had reason to be pleased. Vaccari, for instance, was formerly at Google Maps, as well as the Massachusetts Institute of Technology's (MIT) Senseable City Lab, where he worked on a number of high-profile data visualization projects and coauthored articles on, among other things, engagements with urban space and place as determined from the aggregation of mobile phone activity log data (Girardin et al. 2009). Tretti brought valuable location research expertise of his own, having written a Master of Computer Science thesis at the University of Illinois on the analysis and presentation of results for mobile local search.

The eventual fruit of this harvest of new talent was Nearby. While initially intended as a friend-finder tool (Sottek 2012; Sterling 2013a), by the time of its launch Nearby had been recast as Facebook's "first attempt at local business discovery" and search (Constine 2012f). Nearby provided a "relevancy-sorted list of businesses and landmarks" that Facebook thought each user would be interested in, based on a ranking process that took into account "friends who've Liked a business, checked in, left a short text recommendation, or given the Place a star rating" (Constine 2012f). Each business listing contained category, location, and rating information, and the ability to leave personal tips (Constine 2012f).

The following year, Nearby was split into two parts. The first formed a reinvigorated opt-in friend-finder tool called Facebook Nearby Friends. Built by the Glancee team, Nearby Friends sent users a notification if they were within a short distance of a friend, and if that friend shared his or her precise location, it would appear on a map (Constine 2014). The second was a rebadging of its local business discovery facility as Facebook Local Search (Sterling 2013a). Crucial to Facebook's longer-term vision for Facebook Local Search was the release of Instagram 3.0, which was regeared significantly around the capture and incorporation of location data.

A key aspect of Instagram 3.0 was the introduction of Photo Maps—Instagram's answer to Facebook's timeline—which displayed images arranged by location, as a preferred way of archiving and organizing photo libraries (rather than chronological ordering). In introducing Photo Maps, Instagram CEO Kevin Systrom's larger ambition for geocoded visual data was clear: "We eventually want 100% of photos to be geotagged" (Tsotsis 2012c). Furthermore, he viewed photos not just as a searchable commodity, but as a mechanism for conducting searches, whereby those interacting with Instagram were "using location the same way [one would] explore via hashtags or via a profile" (Tsotsis 2012c, n.p.). What was left unspoken, yet was apparent in these statements, was that there were clear longer-term commercial benefits for Instagram and its parent company Facebook in geocoding pictorial data (Constine 2012a), and it was inevitable that Instagram's datastream would feed into Facebook's—a move signaled by preemptive changes in December 2012 to its privacy policy to accommodate future integration (Crook 2012).

Here it is also worth noting Facebook's US$19 billion purchase in early 2014 of mobile messaging client WhatsApp. A key attraction of this deal was gaining access to WhatsApp's 450 million active users (72% of whom are active each day) (B. Evans 2014). Further motivation for purchasing the service, it has been suggested, was photos: "According to the company's own numbers, WhatsApp is processing 500 million images per day [...]. For its part, Facebook processes a comparatively paltry 350 million photos a day, with an additional 55 million per day from Instagram" (Lacy 2014, n.p.).

All of these moves—the release of Local Search and Nearby Friends, the release of a new version of Instagram, and the acquisition of WhatsApp—collectively mark a significant "ramping up" (Geron 2012) of Facebook's mobile and location ambitions. Data integration with Instagram and WhatsApp, not to mention Friends Nearby, would boost significantly the volume and quality of the geolocation data added to Local Search, and its Graph Search capabilities.

"A location service," Josh Williams once said, "only gets interesting when you get to a certain scale" (cited in Constine 2012c, n.p.). In 2012, it was reported that, each day, Facebook's servers were ingesting 2.5 billion pieces of content, 2.7 billion "Like actions," 300 million photographs, and 500+ terabytes of data, as well as scanning approximately 105 terabytes of data each half-hour (Constine 2012b). At that time—the second quarter of 2012—Facebook reported 552 million daily active users (DAUs) and 955 monthly active users (MAUs) (Facebook 2012). These figures have continued to rise. By the end of 2018, Facebook had 1.52 billion DAUs and 2.32 billion MAUs (Facebook 2019), meaning that the quantities of geocoded and other social media data that Facebook is processing each day are commensurately much greater. Facebook's combination of technical expertise in the area of mobile and ubiquitous computing, the addition of local search, rating, and recommendation functionalities, and the sheer size of its data set have made it a formidable location-based services company.

The net result is that Facebook has now become a major player in mobile location-sensitive social networking, local search, and location-based mobile advertising.

With core location functionality in place, Facebook has continued to make ongoing adjustments and refinements to geolocational offerings across its various services. In 2015, Facebook removed always-on location sharing within its Messenger service—but not before one enterprising person created a Chrome extension that scraped Messenger location data and plotted it on a Harry Potter style "Marauder's Map" of location-tagged messages (Khanna 2015). Taking its inspiration from China's WeChat, which enables users to perform multiple services in-app, Facebook replaced the blue directional arrow with a pin button, "alongside those for sending photos, stickers, or money" (Constine 2015). Tapping this button brings up a map with the user's current location identified by a map pin; tapping it again shares this location with those being messaged (Constine 2015). Then, in 2016, Instagram removed its photo maps, while retaining the ability to view location tags on individual posts (Newton 2016b). The same year, Facebook removed exact location sharing in Nearby Friends, replacing it with a "Wave"—a waving hand emoji that can be sent to someone to signal that they are interested in potentially meeting up (Constine 2016). And, in 2017, WhatsApp launched live location sharing within one-on-one and conversations or group chats (Dua 2017). Characteristic of Facebook's *modus operandi* is the fact that, while it remains crucial to its operations, tinkering with its location offerings is ongoing.

PLATFORMIZATION AND DATAFICATION

In a 2004 reflection on what political economy approaches can bring to the study of "new media," Robin Mansell (2004, 99) writes, "a political economy of new media insists on an examination of the circumstances that give rise to any existing distribution of power and of the consequences for consumers and citizens." Having detailed Foursquare's and Facebook's respective efforts in location capacity building, here I want to give consideration to two more general socio-technical developments—that of platformization and that of datafication. Both have contributed in key ways to the emergence of Foursquare and Facebook as important locative media firms.

As Tarleton Gillespie (2010, 349) defines it, a platform is "an infrastructure that supports the design and use of particular applications." Platformization—"the transformation of social network *sites* into social media *platforms*" (Helmond 2015, 1)—is seen to have become the "dominant infrastructural and economic model of the social web" (1), and has been central to the success of both Foursquare and Facebook. What is significant about platforms is that they "do not seek to internalize their environments through vertical integration"; rather, they are designed "to be extended and elaborated from outside, by other actors, provided that those actors follow certain rules" (Plantin et al. 2016, 298). Thus, "the platform remains a centrally controlled and designed system [. . .], but benefits from the innovations of a large penumbra of third-party developers" (299).

Platforms enable these innovations by making their "services available to other software programs through Application Programming Interfaces (APIs)" (D. Evans, Hagiu, and Schmalensee 2006, vii). In this way, APIs serve as "mediatory objects" in that they not only offer third-party developer access, they also provide a useful way for "companies to extend their reach and growth across the web" (Bucher 2013a, n.p.). What is more, APIs make possible, for firms like Foursquare and Facebook, the "decentralization of data production" on the one hand (by providing "a technological framework for others to build on, geared towards connecting to and thriving on other websites, apps and their data"), and the "recentralization" of data collection on the other hand (by "readying external data for [incorporation within] their own [internal] databases") (Helmond 2015, 8; see also, Gerlitz and Helmond 2013, 1357). Thus, APIs fulfill a crucial function, as Anne Helmond (2015, 8) explains, by creating "data channels"—or what Helmond, following Alan Liu (2004, 59, fn. 22), refers to as "data pours"—for "collecting and formatting external web data to fit the underlying [economic] logic of the platform" (Helmond 2015, 8).

Thus, Foursquare's early embrace of an "open" API has been strategically important and a major contributor to its ongoing viability. Similarly, Facebook's decision to open up its location-related APIs to third-party developers is also interesting in that, at least in the early days of location capacity building, maintaining an "open platform" was a deliberate strategy and certainly assisted it in compensating for lack of first mover advantage in location and mobile by gobbling up geolocation data from elsewhere via its APIs. What Facebook realized, in short, as José van Dijck (2011, 343) has noted in relation to Twitter, is that *geocoded* user data hold greater commercial value for marketers and advertising than non-geocoded data. Thus, as the richness of geocoded Facebook user data increased, it followed that so, too, would the commercial value of these data. As Anne Godlewska and Jason Grek Martin (2011, 365) put it, "the commercial value of the voluminous geographical data being collected daily is considerable."

José van Dijck and Thomas Poell (2013, 9) make the point that, as platforms mature, they turn "more into data firms deriving their business models from their ability to harvest and repurpose data." Not only does rich user-generated geodata mean a platform can set higher advertising rates (Fuchs 2012, 144), but this data becomes the "core, saleable asset" for the owners of the platform (Van Couvering 2011, 198). As Matthew Crain (2018, 98) puts it, "commodification of personal information has become one of the Internet's foremost business models" (see also Andrejevic 2010, 98; Gerlitz and Helmond 2013, 1352; van Dijck 2013b).

It is important to note here that opportunities for commercial exploitation of personal data by these firms have in key respects been made possible by the emergence of data markets and the means for achieving datafication. *Datafication* has been defined as the process by which a phenomenon is transformed into a quantifiable format that can then be analyzed (Mayer-Schönberger and Cukier 2013, 78). Datafication expresses, in other words, the capacity of "commercial digital media [to] capture the details of activity that once eluded systematic forms of value extraction in order to turn them into information commodities" (Andrejevic 2010, 90). In the case of locative media services, what is being captured and analyzed are the traces of our physical and digital passage through space and time, and the "data fumes" (Thatcher 2014) they generate. Our use of locative mobile media mean that our engagement with a specific place or venue forms "one of many shifting data points that define a trajectory" (D. M. Wood 2017, 229), with locative social media analysis, as we saw in the earlier discussion of Foursquare's places and social graphs, targeting people and places simultaneously.

In this way, firms like Foursquare and Facebook have evolved into very sophisticated "profiling machines" (Elmer 2004) that are focused on the "collection, storage, networking, diagnosis, and deployment of demographic and psychographic information" (9). For Carlos Barreneche (2012a), "there is a specific geodemographic ontology underpinning the logic of spatial ordering in location platforms" (337). Platforms like Foursquare and Facebook (and Google, which forms the focus of Barreneche's inquiry) enact "a form of geodemographic profiling that uses the data-mining of records of location-trails to produce the socio-spatial patterns that make up the segmentations that enable making inferences about users' identity and behaviour" (Barreneche 2012a, 339; see also Barreneche 2015, 2012b). Harrison Smith (2017) adds to this understanding of geodemographic profiling by locative media services by noting that

[i]t is also necessary to understand how these processes are being developed beyond the direct perception of users and interfaces, and how the business-to-business aspects of knowing capitalism leverage the production of location data, and the analysis of mobility to produce new kinds of classificatory knowledge, and more importantly, how this knowledge can be integrated with metrics that evaluate the performance of algorithmically distributed content to measure audience response. (H. Smith 2017, n.p.)

Smith labels these processes "second-order geodemographics." What makes them unique, differentiating them from more traditional forms of geodemographic profiling, is that these algorithmic techniques and processes enable specialist location analytics firms like PlaceIQ and Foursquare, as well as Facebook and Google, to understand and produce "observable changes in behaviour," or what is known in the trade as "lift" (H. Smith 2017, 6).

Even more desirous still is the capacity for future-oriented geolocational analysis. Mark Andrejevic and Mark Burdon (2015, 30) argue that "predictive analytics" that "extend into the future"—in pursuit of the desire for "diachronic omniscience" (Parks 2005, 91)—is a key ambition of data-mining firms, including Facebook and Foursquare. Taina Bucher (2012a, 13) observes that Facebook has long been interested in "anticipating attention." And, as noted earlier in this chapter, Foursquare's data scientist Blake Shaw asks, "How can we predict new connections?" This is a question that is made explicit in Foursquare's patent application for its Explore feature, where it explains that

venue preferences may be predicted using such functions in a location-based service based on where the user has previously been, preferences of people in their social network, and/or preferences of certain groups or the entire network of location-based service users. (J. Moore et al. 2013, n.p.)

For locative media firms, "the very capacity to track and predict mobility is directly productive of value" (Barreneche and Wilken 2015, 508). Thus, what we see with Foursquare and Facebook (not to mention Google) are portaits of "user activity made possible by ubiquitous interactivity" that are "increasingly detailed and fine-grained, thanks to an unprecedented ability to capture and store patterns of interaction, movement, transaction, and communication" (Andrejevic 2007, 296). Mark Andrejevic (2007) refers to these as forms of "digital enclosure," in which the developers of location-sensitive mobile, social, and search media services encourage "ever greater participation by the public [that] will be transformed into increasingly exclusive forms of proprietary knowledge" (Andrejevic 2010, 95) for capital accumulation (Kuehn 2016, 3).

CONCLUSION

In the first part of this chapter, I examined the mobile location-based service Foursquare, detailing its still-evolving if slowly solidifying business and revenue models, and documented the company's dramatic strategic change in direction in which it systematically downplayed its gameplay functionality in favor of emphasizing the urban search and recommendation, and location data analytics ("software as a service") aspects of its operations. The preceding factors have prompted Foursquare to rethink its corporate strategy and business model, and the path it has taken—the monetization of user traffic data—is not new. This is not to say that, in a business sense, Foursquare is quite out of the woods yet. Arguably, it continues to occupy a somewhat tenuous position at present within the location services business ecosystem. The difficulty it faces is that, at an industry level, within the "venture capitalist economy of Silicon Valley, partnership deals, even with competitors, are as important as beating the competition" (van Dijck 2013a, 58), and they have proven to be the lifeblood of Foursquare's Places API. Heavy reliance on these arrangements makes them vulnerable to industry upheaval and uncertainty.

In the second part of the chapter, I examined the variety of ways that Facebook cautiously but deliberately set about building geolocation and mobile functionality into its platform offerings. I traced the development and subsequent phasing out of Facebook Places, the extensive background work done after this time to add significant geotagging functionality, the many crucial corporate acquisitions (of Gowalla, Glancee, Instagram, and WhatsApp) that lay the foundations for the next step in geolocation functionality, and the launch and implications of Nearby. Facebook's strategic integration of location into its social media offerings and its emergence as a mobile location-based platform further reinforce the firm's global corporate significance. While Facebook's tinkering with its location offerings is ongoing, it has become clear that Facebook recognizes that the "fourth pillar" of mobile and location (to complement the other three pillars of Facebook's business—its timeline, newsfeed, and search) (Wilken 2014a) is vital to its present and future success. Its redoubling of its efforts around location and mobile have paid significant dividends. Facebook's scale and global reach, and its decision to open its places APIs to developers, has meant the continual enrichment of its "social graph," which, in turn, has substantially strengthened its position as a major player in local and mobile advertiser.

In the final part of the chapter, I suggested that what Foursquare and Facebook have managed to achieve with geolocation has been made possible through platformization and datafication. Facilitated by both of these things, firms like Foursquare and Facebook have evolved into sophisticated "profiling machines" (Elmer 2004) concerned with the collection, transfer, storage, diagnosis, and analysis of vast quantities of end-user-generated geolocation data for capital accumulation. These data collection and analysis processes present manifold challenges, some of which are taken up and explored in Part III of this book. Just as important, though, is the need to balance an account of the economic imperatives and ambitions of locative media firms against careful consideration of the creative as well as routinized quotidian uses of these same services—examination of these concerns forms the explicit focus of Part II to follow.

PART II

Cultures of Use

Introduction

Part II of this book involves a significant shift in focus from what has preceded it. Whereas Part I explored the industry dynamics of locative media, Part II examines cultural practices associated with location-sensitive mobile technologies. Part II is further divided, with the first chapter examining artistic engagement with these technologies, and the following two chapters exploring quotidian uses of these technologies.

Creative engagement with mobile technologies is important to consider here, given the key historical role that locative media arts have played in the development of locative media (Bleeker and Knowlton 2006; Hjorth and Richardson 2014; Leorke 2019; Zeffiro 2012), and given locative media artists' persistent concern for examining the forms of cultural power that shape and are expressed through our day-to-day uses of mobile devices (Martin 2003, 154). The specific works examined in Chapter 4 seek to uncover and reveal the cultural power that is at work in and underpins our day-to-day uses of various forms of locative media, and explores the creative possibilities for critiquing, challenging, and reconfiguring established and emergent power dynamics. I am especially concerned with exploring how much locative media art has been, and continues to be, about the illumination of the tensions of contemporary technologically mediated culture, with art serving as criticism, as an enactment of a subtle political aesthetics. Taking up these themes (as developed by Jacques Rancière), Chapter 4, "Locative Media Arts and Political Aesthetics," explores three specific projects: *You Get Me*, by UK-based art group Blast Theory; *Metadata+*, an iOS app developed by US-based data artist and activist Josh Begley; and, *Border Bumping*, by Berlin-based artist and "critical engineer" Julian Oliver. These projects have been selected for the ways that they utilize different "location" technologies, for the critical issues they raise, and for the opportunities they present for thinking through aesthetics and its relationship to politics.

Turning from the more experimental explorations of locative media arts in Chapter 4, the remaining two chapters in Part II give detailed consideration to everyday cultural practices involving location-sensitive mobile technologies. Chapters 5 and 6 ask: What are the "moods, rhythms and affects" of everyday technology use? And "what are its orchestrations and intensities"? (Highmore 2011, 2)

In Chapter 5, "App Entanglements," I take Foursquare as my point of departure in exploring how use of this particular application routinely occurs alongside and intermingled with the use of a wide array of other applications that also include various forms of location functionality, through a process I refer to as *compartmentalization*.

The claim that location-sensitive mobile social media end-use is characterized by forms of compartmentalization is developed principally from insights generated from an interview-based study of smartphone users conducted in Melbourne, Australia (see Wilken 2015). These Melbourne interviews formed part of a larger comparative project, developed in collaboration with Lee Humphreys of Cornell University, to examine locative media in Melbourne, Australia, and in New York City, United States. This study took place in 2013–2014, and consisted of 31 qualitative, in-depth, semi-structured interviews with locative media end-users in Melbourne and New York City, and four focus groups—two in each city—with small businesses who use locative media. All interviews and focus groups were professionally transcribed, and the analysis of the transcripts was completed using the initial categories that guided the interviews, and then complemented with a thematic analysis to compare responses across interviews. In this book I won't be drawing on the small-business focus group material, as this has been examined elsewhere (Humphreys and Wilken 2015). Interview data are drawn on in the analysis that follows in this chapter, as well as that of Chapters 6 and 8. All participant names have been replaced by pseudonyms, and participant age ranges and occupations have been included.

What emerged from this study of Foursquare end-use was a rich and complicated picture where choices governing the use of and shift between Foursquare and other applications can be seen as shaped by socio-technical dynamics, such as the specific affordances and constraints of each application, and as a result of more finely granulated individual and social practices, such as the recurrent use of Foursquare's check-in history as a personal *aide memoire*, or the segmentation of one's social network and the association of specific groups with specific apps.

Chapter 6, "Territories of the City and the Self: Locative Mobile Social Networking, Urban Exploration, and Identity Performance," again draws from existing research on Foursquare, including that conducted by myself and by others. In drawing from these studies, I examine, first, how urban spaces and places are explored, catalogued, and communicated, and, second, how these communicative practices are entwined with processes of individual identity construction and performance.

CHAPTER 4

Locative Media Arts and Political Aesthetics

The artists of today are grappling with location awareness in the way that much of 20th-century art did with our visual perception of the world.

—Townsend (2006, 347)

[T]o make art is to act.

—Macmillan (1995, 3)

Artworks, then, are part of the general economy of aesthetics in their pedagogic role of alerting us to different kinds of alertness.

—Highmore (2011, 51)

INTRODUCTION: ART AND THE REDISTRIBUTION OF THE SENSIBLE

One of the images that has always stayed with me from my first encounter with the work of the late Scottish concrete poet and artist Ian Hamilton Finlay is that of a garden bird feeder in the form of a miniature aircraft carrier (see Figure 4.1). Reflecting on this and similar pieces, art critic Stephen Scobie (1995, 184) observes that, in Finlay's work, "while the images of the modern Sublime are sometimes presented in their full force, they are equally subject to all sorts of ironic modulation of scale," such as in the stone recreations of aircraft carriers like the *USS Enterprise*, drastically reduced in size, and represented in the form of bird feeders or bird-baths "with little birds instead of warplanes, landing and taking off" (184). These

Figure 4.1. Ian Hamilton Finlay, *Homage to the Villa d'Este*, Stonypath, 1981.
Photograph: John Stathatos. © John Stathatos, 1981, 2017. Reproduced with permission.

artworks speak to Finlay's belief that "certain gardens are described as retreats when they are really attacks" (quoted in Clark 1995, 153).

These uncanny garden ornaments, unsettling in their juxtaposition of the pastoral and the "catastrophic Sublime" (Scobie 1995, 189), were sited within Ian Hamilton Finlay and wife Sue Finlay's garden in the Pentland Hills near Edinburgh. Originally known as Stonypath, this garden was renamed Little Sparta in the early 1980s following a bitter dispute that pitted the Finlays (and their allies, dubbed the "Saint-Just Vigilantes") against the Strathclyde Regional Council over non-payment of rates relating to a particular building within the garden, the Garden Temple of Apollo, and the refusal by the Council to countenance the idea that this building was not (just) an art gallery but in fact a pagan "place of worship" (thus making it, in Hamilton Finlay's view, exempt from rates); this dispute subsequently became known as the "Little Spartan War" (Eyres 1986). Over the course of his artistic career, Finlay has been described as an artist whose work is "militantly political" (Clark 1995, 152). And, yet, what is striking about his poetry and landscape art is the subtlety and wit that often characterize the politics and aesthetics of these works. As Duncan Macmillan notes, throughout his artistic engagement with Little Sparta, Finlay has always used "the garden, the landscape not to irradiate that special aspect of our cultural life which is the perception of nature, but to illuminate, as in a microcosm, the tensions of contemporary culture as a whole" (Bann 1995, 103), thus opening new critical, political, and aesthetic possibilities.

In this chapter, I want to develop a related line of argument to Bann's interpretation of Finlay's art, albeit one that is oriented specifically toward a consideration of locative media art. While not necessarily dealing with landscape (at least not in the sense that Finlay was), much locative media art has been, and continues to be, about the illumination of the tensions of contemporary technologically mediated culture, with art serving as criticism, as an enactment of a subtle political aesthetics. Taking up these themes, this chapter examines three specific projects: Blast Theory's locative game *You Get Me*; Josh Begley's Metadata+ smartphone app; and Julian Oliver's *Border Bumping*. These projects have been selected for the ways that they utilize different "location" technologies (respectively, smartphones and walkie-talkies, GPS, and cell towers); each explores a different set of critical issues that both enrich and complicate our understanding and experience of locative media; and each, in its own way, creates the conditions of possibility—"a pedagogic invitation"—for thinking through aesthetics and its relationship to politics.

For French philosopher and historian Jacques Rancière (2004a), politics and aesthetics and art are brought together by their relationship to a "distribution of the sensible." How Rancière conceives of the interaction between politics and aesthetics requires some explanation, beginning with his quite particular conception of politics. "Against the dominant usage of the term politics that would see it as a theory of parties and policies," Rancière "allows it only one meaning: the enacting of a disruption in the parcelling out of allocated time, space and sense" (Highmore 2011, 47). As Rancière writes:

> Politics is first of all a way of framing, among sensory data, a specific sphere of experience. It is a partition of the sensible, of the visible and the sayable, which allows (or does not allow) some specific data to appear; which allows or does not allow some specific subjects to designate them and speak about them. It is a specific intertwining of ways of being, ways of doing and ways of speaking. (Rancière 2004b, 10)

From this passage, we can see how politics, for Rancière, shapes the "distribution of the sensible." However, "acts of politics and acts of art" can respond to this shaping influence by also working, at times, to "redistribute the sensible," "disrupt[ing] and reorder[ing] the social sensorium, making new experiences possible, making new voices heard," and "altering the horizons of visibility" (Highmore 2011, 48). This can happen, Ben Highmore suggests, in the following ways:

> Dislocations in the distribution of the sensible occur when attention is drawn (again and again) to something that had previously been deemed unworthy of attention, or when someone [who] was deemed as having "nothing to say," speaks in a way that solicits an audience and community of listeners (however small). (Highmore 2011, 48)

As Rancière (2009b, 75) puts it, "what occurs are processes of dissociation: a break in a relationship between sense and sense—between what is seen and what is thought, what is thought and what is felt." As Highmore sees it, Rancière's political aesthetics is most interested in "that initial rip in the distribution of sense":

> [I]n the rip all sorts of things are possible. The rip ushers in emancipation, democracy, equality as its infinite potential. (Highmore 2011, 48)

For Rancière, then, aesthetic experience has a political effect to the extent that it produces (or has the capacity to produce) "a multiplicity of folds and gaps in the fabric of common experience that change the cartography of the perceptible, the thinkable and the feasible" (2009b, 72). As such, aesthetic experience "allows for new modes of political construction of common objects and new possibilities of collective enunciation" (Rancière 2009b, 72; see also, Rancière 2010; Rockhill and Watts 2009).

What is productive in Rancière's work for the present chapter is the understanding that, while "artworks and politics (in Rancière's sense of this term) are joined by their relationship to a distribution and redistribution of the sensible," art can be understood to function at one remove from politics, "preparing the ground for new experiences, and opening up spaces for new subjectifications" (Highmore 2011, 49). Art, in other words, "is always political but its politics is often one that acts as a pedagogic invitation" (49).

Locative media arts—very much a "broadly and diversely constructed field" (Goggin 2006, 202)—has long been at the vanguard (Aceti, Iverson and Sheller 2016; Bleecker and Knowlton 2006; Hemment 2006; Wilken 2012a; Zeffiro 2012) of exploring both the creative possibilities (e.g., Berry 2017; Berry and Goodwin 2013; Berry and Schleser 2014; Hjorth and Richardson 2014; Lemos 2009; Munster 2013; Schleser and Berry 2018; Wark 2002) and critical implications of a wide range of location-enabled mobile, wireless, and remote sensing technologies. Within this field, important work has been done that follows many different trajectories; here I cite just two of these. First, there is critical, creative locative media work examining the "particularities, tensions and conflicts" (Bambozzi 2009) associated with urban space and our movements through it, with the projects often striving to function as "generative displacements" (Rueb 2015) or "dis-locative arts" (Pinder 2013) that work to unsettle our engagements with space, technology, and culture (see, for e.g., Behrendt 2012; Dieter 2015; Farman 2014a, 2014b, 2015a; Frith and Ahern 2014; Galloway 2013; Leorke 2015, 2017; Timeto 2015; Tuters 2012). Second, there is work that also seeks to explore how location-based technologies, including digital maps and the data that feed them (such as GPS, satellite traces, and other data trails), can generate new potentialities for facilitating forms of social appropriation, citizenship, and experimental sociability (e.g., Kurgan 2013; Licoppe and Inada 2006; O'Rourke 2013; Paul 2013; Tarkka 2010; Wilken 2014b), as well as facilitating activist interventions (Duarte 2015; Hudson and Zimmermann 2015; Salmond 2010; Summerhayes 2015; Zeffiro 2015).

The projects examined here—Blast Theory's *You Get Me*, Josh Begley's Metadata+, and Julian Oliver's *Border Bumping*—all continue this rich tradition of locative media art-making and/as critique. As noted earlier, each project has been selected, on the one hand, for the way that it utilizes a distinct, established "location" technology or technologies—smartphones and walkie-talkies in the case of *You Get Me*, GPS in the case of Metadata+, and cell towers in the case of Border Bumping—and, on the other hand, for the way that each project puts these technologies to service in the examination of different sets of critical issues that both enrich and complicate our understanding and experience of LBS. These critical issues range from questions of socioeconomic and cultural difference (*You Get Me*), to meditations on otherness and the brutality and absurdity of armed conflict (Metadata+), to reflections on the arbitrariness and fluidity of national boundaries (*Border Bumping*). What is more, each of these three projects offers, in its own unique way, a "pedagogic invitation" that services a "redistribution of the sensible."

BLAST THEORY, *YOU GET ME*

The first case to be explored in this chapter concerns the work of Blast Theory, an internationally recognized art group based in Brighton, England. Led by Matt Adams, Ju Row Farr, and Nick Tandavanitj, and with a long-standing collaborative relationship with the Mixed Reality Lab at Nottingham University, the group's work, in their own words, "explores interactivity and the social and political aspects of technology" (Blast Theory 2016, n.p.). Many of their projects have sought to pose and explore "important questions about the meaning of interaction" (Blast Theory 2016), both technological and interpersonal, and especially its limitations. One such project, which I concentrate on here, is the locative media game *You Get Me* (2008) (see Figure 4.2). Compared with other Blast Theory projects, *You Get Me* has received little to no critical attention. It has been selected as a case study here for its conceptual richness, and the ways it seeks to utilize various location and communication devices and techniques (mobile phones, walkie-talkies, in-game tracking) to explore themes of distance and proximity, connection and disconnection, and listening and communication across cultures and between classes.

You Get Me was commissioned for Deloitte Ignite, a short festival that ran September 12–14, 2008, at the Royal Opera House, London. The work operates between, and seeks to connect, people occupying two distinct sites that are approximately seven kilometers (or four-and-a-half miles) apart: the London Opera House and Mile End Park, a rejuvenated park that had lain derelict for decades and, in Matt Adams's words, "was symptomatic of the kinds of wounds [that had been inflicted over time] to the East End of London" (pers. comm. 2013).

Figure 4.2. Blast Theory, *You Get Me*, 2008.
Photograph: Blast Theory.

In developing the project, Blast Theory collaborated with eight young people—Jack Abrahams, Hussain Ali, Tendai Chiura, Ivan Neeladoo, Fern Reay, Rita Ribas, Rachel Scurry, and Jade Laurelle Stevens—who all had some connection to Mile End Park (Blast Theory 2008). Each was paid for their involvement—and thus treated as professional collaborators—in the project, which took several weeks to develop prior to its one-off running over the event weekend (Adams, pers. comm. 2013).

Blast Theory worked closely with these eight performers to develop "personal geographies," consisting of "important places or events in their life," which they then formed into a map. In the case of Rita, her map was "arranged around the swimming pool in which she had nearly drowned"; in the case of Hussain, his map became a hybrid landscape: "He decided to have his house in England surrounded by the garden of his relative's house in Bangladesh" (Adams, pers. comm. 2013). Mile End Park was then used as a kind of "biographical stage" (Adams, pers. comm. 2013), or palimpsest, for the overlaying of these personally significant narrative maps. The result was that, as they arrived at the river in Mile End Park, for some of the eight performers, "that would be their bed," while, for others, "that would be the vegetable patch, or whatever" (Adams, pers. comm. 2013). Through this whole process, "critical questions about their lives" emerged for each of the performers, and these questions "came to be the animating force of the work" (Blast Theory 2008).

The game itself—which combined a website, walkie-talkies, and mobile camera phones—functioned as follows. Players of the game logged on to the game site via iMacs stationed inside the Opera House. Upon entering the game site, players were presented with the following text:

> Welcome to You Get Me. This is a game where you decide how far to go. At this moment a group of teenagers are in Mile End Park. Each one has a question they want you to answer. Pick a person and their question. Choose carefully because you only get one shot at this. And the others you didn't choose will then try their best to knock you out. Here they come . . . (Blast Theory 2008)

Every visitor/player, each of whom is given an avatar, must choose from one of the eight performers (known as runners) based on a picture of the runner and his or her question. Hussain was about to move out of home to attend the prestigious Central Saint Martins art school in London, despite parental misgivings about his impending departure, with his parents preferring him to commute from home instead. Hussain's question was: "Why won't your family let you go?" (Adams, pers. comm. 2013). Rachel, as Matt Adams explains, "had this bizarre dramarama kind of lifestyle, where she's telling her boyfriend [. . .] she fancies other people and she feels it's an obligation to be honest about the fact that she's desperately attracted to someone else." Her question was: "What is your line between flirting and cheating?" (Adams, pers. comm. 2013)

After a runner is selected, the player is dropped into the game. The player's first goal is to "listen to the personal geography of [their chosen] runner over the walkie talkie stream"—doing so leads to a deepened understanding of their question. Then the player must navigate his or her avatar through a virtual version of Mile End Park to find his or her chosen runner while avoiding the others; if another runner gets too close, the player is knocked out of the game.

Once the runner is tracked down, the player then types an answer to the runner's question: "If they don't like it, they throw you back: you need to listen to more of their personal geography and come up with a better answer" (Blast Theory 2008, n.p.). However, if the runner approves of, or is intrigued by, the answer, then the medium shifts: the runner switches "to the privacy of a mobile phone" and calls the player; there is also opportunity for text exchanges. As Blast Theory explain, "this one to one exchange allows them to get your direct input into their life" (Blast Theory 2008). At the conclusion of this conversation, the runner then takes and sends a picture to the player. In the case of one "runner," Fern, this read as follows:

> This is Fern. It's 3:45 in the afternoon on Friday 12th September. I'm near the canal with the Pallant Estate behind me and I'm taking a photo for you. You get me. (Blast Theory 2008)

You Get Me builds on Blast Theory's long-standing interest in projects that provide the conditions for and encourage interactions between relative and complete strangers (see Wilken 2014b). This project extends these concerns by making "listening, learning and understanding the core mechanics of the game" (Blast Theory 2008). While geolocation technologies were important to the successful operation of the game, these technical platforms, as Matt Adams explains, were less important than a "staging of distance and proximity and of connection and disconnection" (pers. comm. 2013). These concerns—distance and proximity, connection and disconnection, and listening and communication across cultures and between classes—are captured in the title of the game:

> [T]he title, *You Get Me*, is based on both that idea of understanding someone else and also as a sort of rhetorical kind of device, which has nothing to do with actually asking someone whether they have actually understood what you've just

said. It's a kind of slang thing that has essentially nothing to do with understanding. (Adams, pers. comm. 2013)

Locative games are significant, Larissa Hjorth (2011) argues, in that they lead to transformed understandings and experiences of place and everyday life: they serve to remind us that "places are constructed by an ongoing accumulation of stories, memories and social practices" (84); they encourage a questioning of the "too familiar" routines of everyday life (89); they expose us to "new ways of experiencing place, play and identity" (94) and social interaction; in addition, they can also encourage and involve various modes of listening (K. Crawford 2012). To put this slightly differently, in Rancièrean terms, locative media art and games are part of a "general economy of aesthetics in their pedagogic role of alerting us to different kinds of alertness" (Highmore 2011, 51)—in Blast Theory's works, this is often oriented around a particular "alertness" or sensitivity to others, and to mutual understanding (or its lack). It is an orientation that resonates with philosopher Alphonso Lingis's (1994, 10) concern for being open to otherness and difference—a concern, he argues, which forms through "exposing oneself to the one with whom we have nothing in common." This is worked out in ways that are thoroughly embodied and multisensorial. Lingis writes:

> One exposes oneself to the other . . . not only with one's insights and one's ideas, that they may be contested, but one also exposes the nakedness of one's eyes, one's voice and one's silence, one's empty hands. (Lingis 1994, 11)

Driving this multisensorial, phenomenological engagement with those around us is a belief, for Lingis, that "to recognize the other is to respect the other" (1994, 23). Lingis sees communication as crucial to any exposure to, and recognition and respect of, the other. Communication is understood, here, in an expansive sense to include various technological prostheses *and* various "techniques of the body" (to use Marcel Mauss's formulation). Lingis writes: "We communicate information with spoken utterances, by telephone, with tape recordings, in writing and with printing [. . . and] with body kinesics—with gestures, postures, facial expressions, ways of breathing, sighing, and touching one another" (69).

And, yet, whatever the mode, communication, for Lingis, is always agonistic; that is to say, it is always a "struggle against interference and confusion" (1994, 70). On the one hand, it can function as a "continuation of violence" by other means. On the other hand, and more positively, Lingis argues communication "finds and establishes something in common beneath all contention" (71)—"discussion turns confrontation into interchange" (72). Crucially, though, as Lingis sees it, a key challenge is to recognize that in order to communicate with another, "one first has to have terms with which one communicates with the successive moments of one's experience" (77). These theoretical considerations of language and its possibilities for "interchange" are clearly evident in Blast Theory's locative works. In *You Get Me*, for instance, the gameplay is built around the combination of particular technological prostheses (walkie-talkie, mobile phone, web interface) and particular "techniques of the body" (walking, speaking, sitting, listening) to encourage communicative interaction and interchange with others (game players and runners). Herein lies the real force of Blast Theory's work: the deliberate and systematic creation of "uncomfortable interactions as part of powerful cultural experiences" (Benford et al. 2012) that challenge our understandings of distance and proximity, otherness and identification, and disconnection and connection. Blast Theory's projects, and their probing use of location technologies, provide fertile ground for contemplating and questioning the affordances, as well as routine and wider possible uses of everyday location-enabled mobile communications, as well as their implications for our engagements with those around us.

JOSH BEGLEY, METADATA+

As events unfolded in Iraq in the early months of 1991 during what has since been dubbed the First Gulf War, philosopher Jean Baudrillard wrote a series of essays for the French publication *Libération*, reflecting on and trying to interpret these events. These essays were gathered together and published in France with the provocative title, *La Guerre du Golfe n'as pas eu lieu* (*The Gulf War Did Not Take Place*). In this book, Baudrillard (1995) was not, of course, denying the reality of war—what occurred was a massive aerial bombardment of Iraq's military and civil infrastructure. Rather, what Baudrillard was alluding to in this title was the media spectacle that preceded and accompanied that war. As Baudrillard's translator Paul Patton (1995, 3) explains, "it was not the first time that images of war had appeared on TV screens, but it was the first time that they were relayed 'live' from the battlefront"—these were arguments also developed at length by McKenzie Wark (1994) in his book *Virtual Geographies*. What we saw, Patton writes, "was for the most part a 'clean' war, with lots of pictures of weaponry, including the amazing footage from the nose-cameras of 'smart bombs,' and relatively few images of human casualties, none from the Allied forces"

(Patton 1995, 3). The aerial assault that these nose-cameras documented was in large part made possible by the detailed geolocation data generated by the US military's newly operationalized Global Positioning System, or GPS (Kurgan 2013, 13–14; Patton 1995, 4).

A great deal, of course, has happened since these events, in geopolitical, military-industrial, and governmental terms. We have witnessed, for instance, the rise of al-Qaeda and subsequent rise of Islamic State (IS), the events of September 11, 2001, and numerous other related acts and reprisals since; the controversies that followed WikiLeaks' and Edward Snowden's revelations (see Chapter 7); the introduction of drone ("unmanned aerial vehicle") technologies (see Parks and Kaplan 2017); and further—and now almost constant—conflict in Afghanistan, Iraq, and elsewhere in the Middle East and Africa and beyond.

With respect to the events of September 11, 2001 (or 9/11), the US government responded in very forceful ways, via a series of measures that are by now well known: there was US president George W. Bush's infamous declaration of a "war on terror"; the passing of the US Patriot Act; and, in 2003, the introduction to the US Senate of the controversial proposed Domestic Security Enhancement Act, or Patriot Act II (Michaels 2007).

These (and other) measures have had far-reaching effects and impacts. One of particular relevance to the present discussion is the situation, according to philosopher Giorgio Agamben (2001, n.p.), whereby "security imposes itself as the basic principle of state activity" via a governmental insistence on a lasting (rather than temporary) "state of exception." The declaration of a state of exception imposed by the US government after September 11, 2001, Agamben argues, became "gradually replaced by an unprecedented generalization of the paradigm of security as the normal technique of government" (Agamben 2005, 14). Via ongoing processes of normalization, the emergency ("exceptionalism") transmogrifies and "becomes the rule, and the very distinction between war and peace (and between foreign and civil war) becomes impossible" (Agamben 2005, 22). As Agamben (2005, 30) puts it, the state of exception is resolved into the state of necessity (see also Butler 2004; Neal 2008, 47–48). "After the 9/11 attacks, and with the onset of the global war on terror," Lisa Parks and James Schwoch (2012, 13) write, "all media technologies—whether television newscasts, satellite images, or cell phone videos—were discursively recentered on global security." The situation where exceptionalism has become a permanent state is evident in the extraordinary powers of surveillance granted to, and activities undertaken by, the US National Security Agency (NSA), as revealed by Edward Snowden, and moves by the Obama administration in the United States to open up access of NSA gathered data to other national security agencies "without first applying any privacy protections to them" (Savage 2016). It is also evident in the ongoing, daily use of military pilotless planes—drones—by the US government, and it is this particular form of "locative" technology that forms the specific focus of this second case study.

In light of Baudrillard's arguments about the First Gulf War, what is striking about contemporary military drone use is that there is a complicated interplay between vision and invisibility (Solnit 2014). Camera vision is generated, but it is sourced from military drone recordings rather than the nose-cameras of bombs. Moreover, the "spectacle" of war persists, but this spectacle is no longer televised *en masse*; rather, vision of war is now carefully controlled, and access to this vision is largely confined to drone operators and key US military personnel (Solnit 2014, 304). Given their on-the-ground impacts, and the degree of opacity around the strategic purposes behind their deployment, the lack of public information about drones and drone strikes is, to quote Baudrillard (1995, 81), "a scandal." He writes, "If we do not have practical intelligence about [. . .] war (and none of us has), at least let us have a sceptical intelligence towards it" (58).

"Skeptical intelligence" on drone strikes is precisely what Josh Begley attempts to provide. Begley is a data artist working, at the time of writing, with journalists Jeremy Scahill, Glenn Greenwald, and Laura Poitras, at *The Intercept*, the online publication established and funded by eBay co-founder Pierre Omidyar to report on the documents released by Edward Snowden and to "produce aggressive, adversarial journalism across a wide range of issues, from secrecy, criminal and civil justice abuses and civil liberties violations to media conduct, societal inequality and all forms of financial and political corruption" (Greenwald, Poitras, and Scahill 2014, n.p.). Begley's projects often serve as interventions that make visible "the violence behind the way we live" (Cole 2015, n.p.). In "Officer Involved," for instance, Begley, working under the auspices of *The Intercept*, combined Google Maps' imagery with data gathered by nongovernmental and news organizations—especially *The Guardian*'s "The Counted" project (Guardian 2015)—to document the sites where people were killed by US police in 2015. In writer Teju Cole's introduction to this project, he observes that, in one sense, these innocent-looking aerial and Street View images "are the same as any other stills randomly pulled from Google Maps. But when we look at these photographs in particular, we are also seeing the last thing that some other human being saw. It is an immersion in the environment of someone's last moments." This is what is so disquieting about these images:

They look like insignificant places, but all of them are full of significance for those whose loved ones died there. All are sites of premature death, all are sites where someone was killed, and most also index an unrestituted crime. (Cole 2015, n.p.)

A somewhat similar set of concerns motivates another key project of Begley's, and the one I wish to concentrate on here: a smartphone app that reported US secret drone strikes in foreign territories.

This project is interesting in Rancièrean terms for two reasons: first, for the way that it contributes to a "redistribution of the sensible" by "making new voices heard" and "altering the horizons of visibility" (Highmore 2011, 48). In Begley's case, this includes drawing attention to covert drone strikes and giving a voice—however fleeting—to those impacted by them. In this sense, Begley's endeavors to capture the sites of drone strikes can be regarded as an example of what Maria Miranda refers to as "unsitely aesthetics"—works that create a "disturbance of both sitedness and sightliness" to suggest "a space of tension, ambiguity and potential" (Miranda 2011, n.p.; see also Miranda 2013). Second, this project is interesting for the way that aesthetics functions here, as Rancière (2009a, 13) puts it, as "the thought of the new disorder" whereby "artworks no longer refer to those who commissioned them." With Begley's drones project, unlike the other two projects discussed in this chapter, this "artwork" circulates in key ways independent of its creator by dint of its commercial availability as a downloadable smartphone app that, once downloaded, sits on the end-user's phone and "autonomously" reports drone strike data.

In 2012, Begley developed an iPhone application, Drones+, that would send a push notification every time there was a US drone strike in Pakistan, Yemen, or Somalia to those who downloaded and used the app. Apple rejected the app on five separate occasions, apparently on the basis that it was "not useful or entertaining enough" ("Apple Kills," 2015, n.p.). On the first three occasions, the app was called Drones+; on the last two occasions, it was called dronestream. According to one account, a member of Apple's App Review Team reportedly told Begley that if the app specifically focused on US drone strikes, then "it's not going to be approved. But if you broaden your topic, then we can take another look" (quoted in "Apple Kills" 2015, n.p.).

Heartened by this response, Begley persisted and, with a little "semantic trickery" (Franceschi-Bicchierai 2014), succeeded in having follow-up apps accepted by Apple in 2014. This success was a two-step process. First, he created an "empty" app to test the Apple approval process. The app was called Ephemeral, and its App Store entry consisted of "generic art and text" ("Apple Kills" 2015); it was submitted on January 17, 2014, and was approved five days later. Second, Begley created another app, this one called Metadata+, "which promised real-time updates on national security." A tweet by *Esquire* magazine's Tom Junod was the inspiration for the name of this second app. Junod's tweet read:

> What's the big deal about Metadata? Our government kills people on the basis of Metadata. That's the "signature" in a "signature strike." (quoted in Franceschi-Bicchierai 2014, n.p.)

Six days later, Metadata+ was accepted. Begley then set about (re)populating the app with historical information on US military drone strikes. (Begley apparently kept Ephemeral in the App Store as a back-up should Metadata+ be taken down.) By September 2015, Metadata+ had been downloaded 52,500 times, with most of these downloads coming from the United States, Canada, and Europe (Nath 2016, 317) (see Figure 4.3).

The data populating Drones+, and its successor Metadata+, Begley drew from information amassed by the Bureau of Investigative Journalism, a UK-based independent not-for-profit organization that uses a variety of sources to track and create a record of covert US-led drone strikes. Rather than show images of the aftermath of drone strikes, the messages sent via

Figure 4.3. Screenshot from Josh Begley's Metadata+ smartphone application.
Photograph courtesy of Josh Begley. Reproduced with permission.

the Metadata+ app include a map reference for each strike, and a short text-based description of the human toll of each strike:

> Tariq, 16, and Waheed, 12, were driving to pick up their aunt. A drone ended their journey. (October 31, 2011, Pakistan)
>
> 3 chromite miners—Khastar, Mamrud, and Noorzal—were killed in a car. 4–6 dead, 2 injured. (October 30, 2011, Pakistan)
>
> The second attack of the day killed 4 people in a house. (October 27, 2011, Pakistan)

Metadata+ should also be considered alongside Begley's earlier companion project, *dronestre.am*. This is a web resource and linked Twitter account that also pulls data accumulated by the Bureau of Investigative Journalism to provide details of "every reported covert US drone strike, 2002–2016." Like Metadata+, every *dronestre.am* tweet links a news report of a drone strike with a pithy textual account of that strike:

> Mar 26, 2016: Eight men gathered in a courtyard. Above them, a U.S. drone. Two missiles came out of the sky (Yemen)

The *dronestre.am* project was inspired by conversations Begley had with media theorist and writer Douglas Rushkoff in his New York University Narrative Lab class regarding narrativity and the ways that stories are and can be told on the web. Begley also drew inspiration from artist James Bridle's *Dronestegram* (2012–2015) project (https://www.instagram.com/dronestagram/), which provided documentary, pictorial coverage of drone strikes, as well as from writer Teju Cole's *Small Fates* (2011) project (http://www.tejucole.com/small-fates/), which Begley admired for Cole's poetic engagement with short textual descriptions in the crime-reporting pages of Lagos newspapers that were simultaneously "beautiful and devastating" (Coscarelli 2013, n.p.).

Begley's interconnected projects take as their narrative point of departure, to borrow Lisa Gitelman and Virginia Jackson's (2013, 2) words, the idea that "every move has the potential to count for something, for someone somewhere somehow." In their introduction to the book *"Raw Data" is an Oxymoron*, Gitelman and Jackson suggest that "data need to be imagined *as* data to exist and function as such, and the imagination of data entails an interpretative base" (3): "Data are imagined and enunciated against the seamlessness of phenomena. We call them up out of an otherwise undifferentiated blur" (3). The "interpretative base" that Begley builds from is clearly a very different one from that which provides a foundation for US military usage of this same data. Begley's priority is to "imagine" and "enunciate" the human costs associated with the events that are contained and hinted at within drone strike data. The resultant tweets are striking for their textual brevity. And, yet, this concision belies the power and the pathos of the accumulated content of Metadata+ (a point also noted in Nath 2016, 315). This content seeks to "relocate" (re-site) events that the abstractions of military-speak and lack of wider press coverage mean are "dislocated" (unsited because unsightly). That is to say that Begley's project "relocates" in the sense that it places the victims in particular times and places, thereby bringing some context to their deaths. This is done regardless of the victims involved. For instance, the drone strike that performed a "targeted killing" of "top Islamist militant commander" Jalal Baleedi, who is said to have led al-Qaeda's combat operations in southern Yemen, and was reputed to have defected to IS, is reported with the same even hand as unintended targets of drone strikes—or what the US military calls EKIA, "enemies killed in action" (Begley 2015):

> Feb 4, 2016: Jalal was driving through coastal Abyan. A drone incinerated him from above. 2–3 others killed. (Yemen)

While the *dronestre.am* Twitter account and associated website are still operational and continue to report on drone strikes, in 2015 Apple pulled Metadata+ from its App Store. Citing provision 16.1 of its App Store guidelines, Apple informed Begley that this was done on the grounds that the app contained "excessively crude or objectionable content" (Wingfield 2012). Following press coverage of the Metadata+ case, Apple also subsequently withdrew Ephemeral from the App Store.

This banning of Metadata+ led to criticism regarding the vetoing and gatekeeping role played by Apple, and the implications that in so doing it was appeasing the interests of US government security agencies. And, yet, Begley maintains that these App Store struggles were and are not the most important aspect of these interlinked projects. In the case of Metadata+, "the app is not about Apple or Android or the process for approving apps," Begley argues. Rather, "it's about tracking and mapping covert war" (Begley, quoted in Franceschi-Bicchierai 2014, n.p.).

Anjali Nath (2016, 319) asks, "how does remote witnessing shape social and political practice, and relations between people?" In seeking to answer this question, Nath observes that Begley's applications "can hardly be considered an intervention that affects the everyday functioning of drones" (320), nor are they "of immediate use to the actual populations vulnerable to these strikes" (326). These are not what the project was intended to do. Rather, Begley's apps were "specifically designed to bring these [US-government led] violent attacks into view within the American public sphere" (326). For Nath, while "people who download the [Drones+ or Metadata+] app presumably have their own strong opinions about drone violence already" (327), Begley's work is striking "as a critical effort that brings into focus how quotidian smartphone interfaces [can be used to] bear traces of distant wars" (327). Thus, what is uncomfortable and so compelling about Begley's work is that it offers, on the one hand, a critique of opacity (the lack of scrutiny of "extrajudicial killings"), and, on the other hand, "a critique of objectivity" (Gitelman and Jackson 2013, 4) (the assumption that the geolocated SIM card or mobile handset and other data used to order a drone strike are accurate and that the "objective" of a drone strike is an "intended target" and not, as is so often the case, an "enemy killed in action," a.k.a. an innocent victim). In Begley's own words, his aim for the *dronestre.am* website and associated @dronestream Twitter account and for the Metadata+ app was to create "a living archive of hauntings— those which ghost the landscapes we create, and those which ghost the landscapes some of us will never have to see" (Begley, quoted in Franceschi-Bicchierai 2014, n.p.).

JULIAN OLIVER, *BORDER BUMPING*

Julian Oliver is a New Zealand–born, Berlin-based media artist, or "critical engineer" (Oliver, Savičić, and Vasiliev 2011). Oliver has had a diverse creative career that has taken him from architecture via computer games to media art/"critical engineering." His artistic practice, as Taina Bucher (2011, n.p.) explains, "clearly reflects his hacker and gaming background" and includes "playing around and messing with routers, capturing data from open wireless networks, visually augmenting commercial billboards in the cityscape, sonifying Facebook chats, visualizing protocols and otherwise manipulating networks for artistic purposes." Oliver's art is very much activist art. Indeed, as he puts it in a 2016 tweet: "Art that isn't activism decorates" (Oliver 2016, n.p.).

The project examined here, *Border Bumping* (2012–2014), was commissioned by the Liverpool-based Abandon Normal Devices (AND) festival for their 2012 Mobile Republic program, where car-towed "touring caravans"—"a symbol of the traditional British summer holiday"—were "remodelled and re-imagined by artists, architects and activists" (http:// www.andfestival.org.uk/events/mobile-republic-digital-caravans-2012/). Oliver pitched *Border Bumping*, a project that examines the "contradictions of territory when cellular infrastructure from one country bleeds over the border and into another" (Oliver 2014b, n.p.); *Border Bumping* was subsequently further developed during a short residency Oliver held in 2012 at the Technē Institute at the University of Buffalo, New York.

Central to *Border Bumping* is a critical engagement with cellular telephony infrastructure. A crucial component of this infrastructure is the Base Transceiver Station (BST), those bulky, often roof- or pole-mounted installations of antennae, amplifiers, transceivers, and other hardware componentry. As Oliver explains,

> Every cellular BTS has three primary identifiers: a Mobile Network Code (or MNC), a 2 or 3-digit number identifying the tower operator; a Mobile Country Code (or MCC), a 3-digital number indicating the country the BTS is licensed to operate within; and the Cell ID, a unique and longer number identifying the BTS itself, wherever it is. The MNC is used by phones to determine if the BTS is one they're allowed to register with (home network or roaming partner) whereas the MCC denotes the country in which the BTS resides—visible most notably upon receiving the welcome message when crossing a border. (Oliver 2014b, n.p.)

Within this particular project, Oliver was primarily interested in the second of these identifiers—the MCC—and the process whereby a user of a smartphone in close geographical proximity to the borders of two countries finds their phone registering with a BTS on the other side of the border, "reporting itself [via the MCC] to be in that country without actually being there" (Oliver 2014b, n.p.). As Helga Tawil-Souri (2015, 160–161) notes, "cellular signals by their nature do not 'know' to stop at political boundaries."

The *Border Bumping* project consisted of four component parts. The first involved Oliver "engaging in extensive field research at a national border" (Oliver 2014a, n.p.), systematically sweeping, using a "110cm directional antenna reminiscent of a fish skeleton" (Oliver 2014b, n.p.), for cell towers—especially "stealth cell towers," those disguised as other objects—and noting their "unique CellID, signal strength, country code and network code (owner)" (Oliver 2014a) (see Figure 4.4).

The second component part, undertaken in collaboration with Till Nagel and Christopher Pietsch, involved collating this data to create annotated "tele-cartographic" maps that displayed the "borders as they are 'bumped' by border crossing events" (Oliver 2014a, n.p.); the map also displayed other information sent from the users, such as the cell tower ID, signal strength, and so on. The central conceit of the *Border Bumping* project is found in the next step: the redrawing of national boundaries on these maps

Figure 4.4. Julian Oliver, *Border Bumping*. Sweeping for cell towers.
Photograph by Matt McCormick; photograph supplied by Julian Oliver. Reproduced with permission.

to reflect not existing geopolitical boundaries of nation-states, but, rather, "reality [as understood] from the cellular networking perspective" (Oliver 2014b). In Figure 4.5, we see these redrawn national boundaries as jagged black lines that have departed from existing geopolitical borders to meet the small black dots on the edges of the two larger red circles, which correspond with BTSs previously in another country.

The third component of the project was a "stealth cell tower archive," developed by Chris Pinchen, with a web interface so that the public could discover, or contribute further information on, the location of disguised cell tower infrastructure. While the fourth and final component of the *Border Bumping* project was an Android smartphone application that was used to record border bumping events, or what Oliver (2014b, n.p.) refers to as those "moments of bleed or discrepancy between network and national territories."

To give an example of *Border Bumping* in action, an individual might approach the national border separating Germany from France with the *Border Bumping* smartphone app running. While she might be in Germany, her device reports that she is in France. The app prompts her to log this "incident" (see, for example, Figure 4.6). Once sent to the server, this report is then taken literally and, on the accompanying map, "the French border is redrawn accordingly" (Oliver 2014a, n.p.). As Oliver (2014a)

Figure 4.5. Julian Oliver, *Border Bumping*. National boundaries depart from existing geopolitical borders and are redrawn as jagged black lines that meet Base Transceiver Stations (BTSs) that were previously in another country.
Source: Julian Oliver. Reproduced with permission.

puts it, "The ongoing collection and rendering of these disparities results in an ever evolving record of infrastructurally antagonised territory, a tele-cartography."

Oliver considers *Border Bumping* a "work of dislocative media," one that "situates cellular telecommunications infrastructure as a disruptive force, challenging the integrity of national borders" (Oliver 2014a). David Morley writes of how "communication technologies can function as disembedding mechanisms, powerfully enabling individuals [. . .] to escape [. . .] from their geographical locations" (2000, 149–150). With *Border Bumping*, however, the process is less an *escape from* geographical locations as it is a *reinscription of* them along infrastructural rather than geopolitical lines. The significance of *Border Bumping* resides in the way that it exploits cell tower recognition "errors" to not only "question the integrity of national borders" (Forlano 2013, n.p.), but to then exploit these "errors" to redraw otherwise accepted maps in order "to account for the mediation processes of borders" (Boyle 2015, 18).

Oliver's project is symbolically important for the intervention it makes into ongoing debate concerning Europe's internal and external borders

Figure 4.6. Julian Oliver, *Border Bumping*. The *Border Bumping* app prompts the user to log an "incident."
Source: Julian Oliver. Reproduced with permission.

(see, for example, Balibar 1998; Massey 2005; Morley 2000). Despite the 1985 Schengen Agreement to permit greater freedom of movement within Europe (Maas 2013), the delineation of national borders remains of vital importance within Europe (as elsewhere). As Katja Aas (2013) points out, the clear delineation of territory is of vital importance "as a material as well as symbolic expression of the nation state's ordering capacity and national identity" (195–196). And, as we have seen with European responses to the ongoing Syrian refugee crisis, they are also important in (controversially) enforcing a sense of "fortress Europe." Yet, maintaining freedom of movement (of people, of goods, of information and communication) remains a vital concern. Thus, as Aas (2013, 199) points out, "contemporary governments seem to be caught between two contradicting impulses: on the one hand, the urge towards increasing securitization of borders, and on the other hand, the awareness of the importance of global flows for sustaining the present world economic order." Writing on these boundary issues, geographer Doreen Massey suggests that

> [t]he real socio-political question concerns less, perhaps, the degree of open-ness/closure (and the consequent question of how on earth one might even begin to measure it), than the *terms* on which that openness/closure is es-tablished. Against what are boundaries established? What are the relations within which the attempt to deny (and admit) entry is carried out? What are the power-geometries here; and do they demand a political response? (Massey 2005, 179)

The importance of the issues Massey identifies, and of the forms of playful infrastructural critique embodied in Oliver's project, is made abundantly clear in Helga Tawil-Souri's (2015) account of the struggles over mobile te-lephony access and infrastructure between Israel and Palestine. "In Israel/Palestine," she writes, "telecommunications infrastructure is not a meta-phor for the conflict, it is the conflict in material form":

> Who can call what number on what network from what location and at what price are deeply political concerns shaped by the uneven relationship Palestinians and Israelis have to the construction and enforcement of territorial borders. (Tawil-Souri 2015, 157)

For Israelis, they are serviced by near blanket coverage due to the fact that "*all four* Israeli providers have dozens of antennas, transmission sta-tions, and additional infrastructure across occupied territory," and, in the Migron outpost, the contested placement of cell towers "serves as the roots for [further] territorial colonization" (160). For Palestinians, in contrast, "making *any* kind of telephone call" is difficult due to tight legal and military restrictions governing telecommunications systems (162) that constrain access and infrastructural development. The result, Tawil-Souri writes, is a situation where "landline telephone networks, cellular networks, and Internet connections invoke the parcelization and fragmentation of the Territories themselves" (165).

Both Oliver's *Border Bumping* (dis-)locative media art project and Tawil-Souri's account of Israeli/Palestinian struggles over communications clearly draw out the ways that "media infrastructures and networks—such as telecommunications—are not in and of themselves boundless and open but function as politically defined territorial spaces of control and are integral aspects of states' *territoriality*" (Tawil-Souri 2015, 158). These infrastructures are "outcomes of social, economic, political, and territorial processes," and, as such, raise important questions concerning what Doreen Massey (2005, 181) would term "a politics of connectivity."

CONCLUSION: "VIOLENCE," OR, THE "RIP" IN THE DISTRIBUTION OF SENSE

This chapter opened with the artwork of Ian Hamilton Finlay; I wish to close the chapter by returning to his work and to the playful juxtaposition of neoclassical imagery and themes with more recent (especially military) iconography that is central to his often-pointed critiques of the modern condition. Even within the more overt—and seemingly obvious—of these juxtapositions, there is a subtlety, wit, and depth of critique that belies (the more obvious) surface appearances. Take, for example, *Arcadia*, a 1973 green screen print on paper of a camouflaged Panzer tank, produced in collaboration with George Oliver (see Figure 4.7).

There are two clear artistic allusions at play in this work. The first of these is a reference to a pair of paintings by seventeenth-century French artist Nicholas Poussin, entitled *Et in Arcadia Ego*, which depict a group of shepherds discovering a tomb. The title of Poussin's works are drawn from

ARCADIA

IAN HAMILTON FINLAY/GEORGE OLIVER · WILD HAWTHORN PRESS

Figure 4.7. *Arcadia* [collaboration with George Oliver], 1973; Ian Hamilton Finlay (1925–2006); Medium: lithograph.
Source: Tate, purchased 1974; © Estate of Ian Hamilton Finlay; Photo: © Tate, London, 2017. Reproduced with permission.

Virgil's *Eclogues* V 42, and both sets of artworks—those by Poussin and that by Finlay/Oliver—operate as a *memento mori*, a reminder that death is present everywhere, even in paradise. The second, as Nicholas Davey (2012, 25) points out, concerns the visual devices Finlay/Oliver use, where the choice of forest green, the camouflage foliage (which evokes "Arcadia's glades"), and the form of the image ("its simplicity, detail, and location on the page" and typographic style) are all "reminiscent of late eighteenth century folios of flora and fauna illustrations." These associations—and the placement of "the image of a military vehicle into various fields of classical meaning"—combine, Davey argues, in ways that prompt us to rethink our existing understandings of Arcadia and, I would add, to reflect on the implications of our contemporary acceptance of the military-industrial complex. Here "the speculative charge of word and image which the 'force' of Finlay's work relies on, pushes into deeper and, perhaps, more sinister complexes of meaning and association" (Davey 2012, 26).

A more recent artistic reworking of the same military hardware, albeit explored in pursuit of a somewhat different set of critical concerns, is found in Julian Oliver's *No Network* from 2013 (see Figure 4.8). Here Oliver has taken a 1:25 scale model of a 1966 British Chieftain tank, which he has then outfitted with a jammer board and an antenna in place of a gun. When activated via the tank-top switch, the *No Network* tank jams a variety of mobile telephony signals (including calls, SMS, and 3G data packet networking)

Figure 4.8. Julian Oliver, *No Network*, 2013.
Source: Julian Oliver. Reproduced with permission.

that are in operation in Europe, North America, and Asia within a range of 6–15 meters.

Additional tanks are in development that carry other geolocational "weapons" capabilities, including those that will block "GPS location services and 802.11 (WiFi) wireless networking" (Oliver 2013). "In an age of mass communication," Oliver (2013, n.p.) writes, "the right to not be contactable, to be 'off-the-grid', is enforced in the company of this object." Thus, if this artwork functions as a *memento mori*, it is in a very different sense from those artworks discussed earlier: here *No Network* serves to remind us that "signal death" is ever present, but that, in a "networked" age, to view communications technology as "both a challenge and a threat" (Oliver, Savičić, and Vasiliev 2011, n.p.), as Oliver does, and to strive to deliberately achieve signal death, thereby interfering with corporate and government interests and hindering desires for seamless connectivity, is nothing short of a provocative act and will itself be considered a challenge and a threat.

A crucial point to take from these final two projects and the others explored in this chapter is well drawn out by Davey (2012, 26) when he writes: "As the Ancients knew, art marks on the one hand the collision of the forces of order and discipline with the disruptive energies of violence and disorder on the other." This is what we have also come to know, via the writings of Rancière, as the *redistribution of the sensible*. As Ben Highmore (2011, 48) points out, "violence" in Rancièrean political aesthetics has a quite particular meaning: it refers to, and is most interested in, "that initial rip in the distribution of sense," those aesthetic experiences that are political insofar as they create "a multiplicity of folds and gaps in the fabric of common experience that change the cartography of the perceptible, the thinkable and the feasible" (Rancière 2009b, 72). In the final two examples, the incorporation of military machinery plays a vital part in creating the initial "rip" that can lead to a critical re-exploration of possibilities beyond the proliferation of military weaponry and power and its environmental impacts, on the one hand, and the assumed normality of and desire for ubiquitous, unbroken telecommunications connectivity, on the other. As this chapter has sought to demonstrate, the redistribution of the sensible can occur in manifold, powerful ways within locative media arts. In the first of the three cases explored in this chapter, a simple yet very personal question constituted the "rip" that led to communicative interactions at a distance that sought to open up new possibilities for engagement with (and the creation of intimacy between) strangers. In the second case, the creation of an iPhone app—and the stream of information flowing through it—sought to become (but was repeatedly hindered from becoming) "a living archive of hauntings" that might produce a "rip" by using "the very technologies that produce the world as a place of transnational digital neoliberal consumption, to illuminate the destruction of entire life-worlds, the production of other spaces and people as constantly under threat, and the persistence of sonic disruption and surveillance" (Nath 2016, 327; see also Kafer 2016, 65). In the third and final case, the somewhat innocuous occurrence of a cell phone signal being detected by a Base Transceiver Station in another country became the "rip" that permitted the possibility of reconfiguring geopolitical boundaries along different lines and thus of reconceiving the nation.

The remaining chapters in this section on the "Cultures of Use" of locative media take a very different angle of approach from the present one. Over the next two chapters, the focus of the book shifts from the more experimental explorations of political aesthetics within locative media arts, to considerations of consumer end-users and end-uses of commercially oriented location-based search and recommendation and social media smartphone applications. In the next chapter, I explore the ways that use of any smartphone application routinely occurs alongside and as intermingled with the use of a wide array of other applications that also include various forms of location functionality.

CHAPTER 5
App Entanglements

My Instagram is definitely my expression and what I love to show the world. My Snapchat is kinda my silly sarcastic alter ego. . . . Twitter is where I can freely talk and have conversations with anyone and everyone! I feel such a connection on Twitter.

—Kardashian West (2016)

How I organise my messaging:

SMS: People I see regularly

Line: People who speak Japanese

WhatsApp: Asian people I'm related to and people in Southeast Asia

Facebook Messenger: People I mainly hung out with from 2005–2008 and white people I'm related to

—Liaw (2018)

I use different social media depending on who—obviously, for me, I've kind of *compartmentalised* them.

—Melbourne Foursquare user, Mario (aged 35–44, marketing; emphasis added)

INTRODUCTION

"To classify is human," write Geoffrey Bowker and Susan Leigh Star (2000, 1). "We spend large parts of our days doing classification work, often tacitly, and we make up and use a range of ad hoc classifications to do so. We sort dirty dishes from clean, white laundry from colorfast, important email to be answered from e-junk" (1–2). Bowker and Star's point is that, however integrated into our daily lives, and thus ordinarily hidden from view, everyone uses and creates classifications in some form each and every day (6). This is because, like lists (Wilken and McCosker 2012a, 2012b), classificatory categories "simplify the world into meaningful and useful chunks" (Bhaskar 2016, 161) that "do some kind of work" (Bowker and Star 2000, 10) for us. Writing on the pioneering efforts of Swedish botanist Carl Linnaeus, Michael Bhaskar (2016, 160) notes that Linnaean taxonomy reminds us of the importance of classificatory categorization: "without categories, we struggle." "Every area of knowledge and activity," Bhaskar (2016, 160–161) writes, "requires a prior process of categorisation." We also rely on categorization processes for the deceptively simple reason that one of their key purposes is to make the exchange of information easier.

Building on these understandings, this chapter advances the claim that there is a tendency toward classification that involves forms of *compartmentalization* in the use of location-enabled mobile social networking services that shape app selection (which app is selected and for what purposes, based on how one perceives the affordances or utility of specific smartphone applications), and app use (especially in relation to what is shared through specific apps and with whom). What becomes evident, I argue, is a rich and complicated picture where choices governing the uses of, and shifts between, various applications—a dynamic Carolyn Haythornthwaite (2005, 130) refers to as "media multiplexity"—can be seen as a result of finely granulated individual and social practices around the management of aspects of one's social networks and app use, and as a result of shaping by socio-technical dynamics, such as the specific "affordances" and constraints of each application or platform. Understanding the complexities driving these end-user choices of application selection and interaction is crucial if we are to more fully grasp the ties that bind platforms, political economies, and publics, and if we are to critically respond to key policy considerations, such as the privacy impacts and implications of location-based services (which will form the focus of Chapter 8 of this book).

In exploring app end-use, I preference the notion of *compartmentalization*— the act of organizing or arranging "into compartments or categories" (Hughes, Michell, and Ramson 1992, 222)—over other, related terms, such as "tiering" (Dena 2008), "boundary work" (boyd and Marwick 2011), and the "dialectics" of friendship and disclosure (Petronio 2002; Rawlins 1992), "segmentation" (Davis and Jurgenson 2013), or "segmentarity" (McCosker 2017a, 2017b), for a number of reasons. First, what is particularly suggestive about compartmentalization (and is less apparent in these other terms) is that it conveys the twin processes of *differentiation* and *categorization* that occur across end-user engagements with social media services. Second, while "segmentation" and "segmentarity" reference the quite deliberate ways that social media end-users and the data they generate are "segmented" by platform owners and marketers along demographic (and other) lines for commercial exploitation (issues that have been addressed in Chapter 3), here the focus is principally on the agency of end-users in making decisions about how and why they select and post to location-sensitive apps. Third, compartmentalization carries within it an understanding of "compartment" as "an area of activity etc. kept apart from others in a person's mind" (Hughes, Michell and Ramson 1992, 222); that compartmentalization processes are defined as principally mental operations will prove important for the discussion to follow. And, building on this third point, the fourth reason is that the idea of compartmentalization as it is intended here dovetails with psychological understandings of compartmentalization, which "allow for contextualized multiple selves" that are "actively constructed by individuals who shape their own self-categories and contexts as part of their general motivation to make sense of the world and to function adaptively within it" (Showers and Zeigler-Hill 2007, 1182; see also Showers 2002; van der Nagel 2018).

The claim that location-sensitive mobile social media end-use is characterized by forms of compartmentalization is developed principally from insights generated from the interview-based study of smartphone users conducted in Melbourne, Australia, and New York City, that has been described in the Introduction to Part II (see also Wilken 2015).

A key aim of the interview component of that study was to understand how people engaged with and used a specific locative media app (Foursquare) alongside and as well as a wider suite of location-sensitive mobile social networking apps (Facebook, Twitter, etc.). Thus, informing the study was an understanding that location-enabled apps operate within a wider "ecology of communication," as understood in David Altheide's precise sense of this term, and specifically as set out in his 1994 article in *The Sociological Quarterly*. Altheide (1994, 667) argues that there are three dimensions to an ecology of communication: an information technology; a communication format; and, a social activity. Not only does Altheide's use of the term "format" clearly reference the complicated history of competing technology formats—such as the "format wars" between Betamax and VHS (Benson-Allott 2013; Greenberg 2008; Wasser 2001)—it also bears a marked resemblance to what contemporary internet researchers refer to as a social media "platform" (Gillespie 2010). Altheide (1994, 668) writes: "Every medium of communication and the information technologies used to shape and transmit information does [sic] this through certain patterns, shapes, looks and these we refer to as formats." Formats, Altheide adds, "provide the basic meaning pattern to an activity" (668); formats "structure the purposes [. . .] to which ITs apply" (668).

In the context of this chapter, a communication ecology approach thus combines information technology (mobile devices), communication format (location-sensitive social media platforms and associated applications), and social activity (social networking).

For Altheide, there is fourfold merit in adopting an ecology of communication approach: (a) it "implies relationships [are] related through *process and interaction*"; (b) it "implies a spatial and relational basis for a subject matter"; (c) it suggests that "relations are not haphazard or wholly arbitrary" and that "connections have emerged that are fundamental for the medium (technology) to exist and operate as it does"; and, (d) it reveals that "there are developmental, contingent and emergent features of ecology," which is to say that an "ecology does not exist as a 'thing,' but is a fluid structure" (Altheide 1994, 667) that changes over time. In combination, these things are "intended to help us understand how social activities are organized and the implications for social order" (667).

A communicative ecologies approach not only informed the study being reported on here, it also constitutes a key framework for this book as a whole that, when applied at different scales, provides a productive means of grasping the workings, interactions, and significances of locative media and wider location-based services. At a macro level, it helps us make sense of the variety of interconnections and interactions that are vital to the formation and successful operation of the larger location-based services, or geo-services, industries (Part I, Chapter 1). At a meso level, this approach assists us in understanding the corporate arrangements that give shape to the mapping, search, and social media businesses that generate geodata and rely on geolocation functionality (Part I, Chapters 2–3; Part III, Chapter 7). And, in the present chapter and those that follow (Part II, Chapter 6; Part III, Chapter 8), this approach is productive at a more micro level for understanding consumer end-uses of a variety of location-enabled social media smartphone applications. Thus, in Part I of this book I argued that it is vital for us to consider the political economies of search and social media companies; in Part II, I make the claim that it is equally vital that we consider everyday consumer end-usage of these same apps and services (McCosker and Wilken 2017). In short, the argument of Part II of this book is that "practices matter" (Couldry 2012; McCosker 2017a). As José van Dijck (2013a, 6) points out, *usage* is a very different concern from (imagined) *users*. By thinking about *usage*, we are able to get at what Heidi Rae Cooley (2004, 148) terms "the intensity of particularity." What is it, in short, that people *do* with both specific location-based social network apps and with their larger suite of installed apps?

COMPARTMENTALIZATION AND APP UTILITY

One of the preliminary questions asked of the Melbournians I interviewed was what location-based and other location-sensitive mobile social networking applications they used. While the search and recommendation service Foursquare formed the initial point of departure for discussion in interviews, it was but one of a larger suite of social networking apps these participants had access to that relied upon and produced various forms of location data. In addition to Foursquare, others that were explicitly mentioned by participants included well-known services, such as Facebook and Instagram, Twitter, Snapchat, Pinterest, Waze, Google+, the defunct Google Latitude, and Apple's Find My Friends, as well as less well-known services, including GetGlue (a social media service for TV fans, which became tvtag before closing in 2014), Twinkle (a now discontinued location-aware client for Twitter), social media service Path (purchased by Kakao Corp. in 2015), and Runkeeper (acquired by ASICS in 2016) and Strava, these last two being location-based services specifically for runners and cyclists.

One of the striking aspects of the responses to this question, illustrated in the quotes that follow, was how each respondent had a pre-established and very clear sense of what they perceived the main purpose or utility of each of the applications they listed to be.

> Foursquare is what I use purely for checking in and finding out about actual locations and suggestions and things like that. [. . .] Facebook, that's how I keep in touch with a lot of friends of mine. Twitter is . . . the main thing is the information that I get from it in terms of news, opinions and sort of current events, all that sort of stuff, and humour I guess as well. (Lachlan, media production, aged 25–34)

> I sort of like Tumblr and Instagram for the visual elements and I like that Twitter is succinct. So, I guess I like that instantaneous kind of information, like, whether that's visual or a sentence or so. I had a two-year hiatus [from Facebook] and now I only use it to promote things that I wouldn't be able to otherwise on other sorts of platforms. (Lucy, student, aged 16–24)

> I've got Twitter to check news and I have some friends on Twitter as well. Basically, everything that's on Facebook, I use on different applications. Like, for updates I have Twitter, for pictures there's Instagram; I don't like how Facebook has all your information. (Mary, executive assistant, aged 25–34)

A number of things are striking about the preceding insights. First, they highlight the observation that social media tools "enable the possibility for more controlled and more imaginative performances of identity online" (Papacharissi 2010, 307); this is an issue that will be taken up and explored in detail in the next chapter. Second, they illustrate vividly the point that "every participant in a communicative act has an *imagined audience*" (Marwick and boyd 2011, 115; for detailed discussion, see Litt and Hargittai 2016). Third, the preceding passages register responses that largely accord with how these services were intended to be used by their designers—that is to say, none records unintended or "off-label" uses (Albury and Byron 2016; Bercovici 2014); nevertheless, the clear demarcation between, and compartmentalization of, each service is particularly striking (see, also Özkul 2015a, 111).

A valuable way of making critical sense of the second and third of the preceding points is by turning to the notion of "affordance," and especially Peter Nagy and Gina Neff's (2015) work on "imagined affordances." "Affordance" is a multivalent concept, and is one that has developed according to a rich "intellectual trajectory" (Bucher and Helmond 2018, 235). It emerged first within ecological psychology (J. Gibson 2015), with subsequent take-up within technology and design studies (Norman 1988, 1990), human-computer interaction (Gaver 1991), and, later, within sociology, and media and communication.

Within the humanities and social sciences, affordance has become "a key term for understanding and analysing social media interfaces and the relations between technology and its users" (Bucher and Helmond 2018, 235). It forms an important concept because it provides a means of accounting for the ways that technologies "are socially constructed on the one hand, and materially constraining and enabling on the other hand" (Bucher and Helmond 2018, 238). And, yet, even within the disciplinary context of greatest interest to me here—media and communication scholarship on social media use—there is considerable variation in how the concept of affordances is understood and applied analytically. Tracing this complex of conceptualizations and uses is not my immediate concern (for detailed discussion, see Bucher and Helmond 2018; S. Evans et al. 2017; Nagy and Neff 2015; Schrock 2015); rather, my focus here in engaging with this literature is to draw specific insights that productively shed light on people's *perceptions* of the utility of specific apps. In this respect, Nagy and Neff's formulation of "imagined affordances" is especially valuable for the reason that the issue of end-user expectations is central:

> Users may have certain expectations about their communication technologies, data, and media that, in effect and practice, shape how they approach them and what actions they think are suggested. These expectations [may or] may not be

encoded hard and fast into such tools by design, but they nevertheless become part of the users' perceptions of what actions are available to them. (Nagy and Neff 2015, 5)

As we saw in the earlier passages by Lachlan, Lucy, and Mary, their expectations or perceptions of what certain platforms were to be used for were in fact "encoded" into these tools "by design." This is not always the case, however.

Imagined affordances, Nagy and Neff (2015) argue, tend to "emerge between users' perceptions, attitudes, and expectations; between the materiality and functionality of technologies; and between the intentions and perceptions of designers" (5). This is to say that what an end-user *believes* he or she will or won't do, or can or can't do, with certain technological tools is "mediated and shaped in relationship to perception, imagination, design, and use" (6). While Twitter would undoubtedly prefer that we *do* geocode our tweets (Wilken 2014c), for Marcus this does not accord with his perception of how the platform should be used: "People on Twitter don't care where you are, if they did, they'd follow you on Foursquare. To me, Twitter's not about location or where you are, it's more just putting up your funny line or your comment" (Marcus, media production, aged 25–34). Meanwhile, Mario (marketing, aged 35–44) describes how Foursquare, prior to the splitting-off of its social features into the Swarm app, was for him "not about a network connection as such, but more about a record of where I've been, maybe where I'd like to go to." Mario also noted that "the value for me of Waze is not about people," despite it being promoted at the time as a social network of sorts; rather, "it is about crowd sourced traffic." And, to give a somewhat different example, Caroline (executive assistant, aged 25–34) expressed frustration at what she perceived she *couldn't* do on a particular platform. Caroline admitted to pushing all her Foursquare check-ins to Twitter, and checking in separately via Facebook at a venue as well. This, she explains, is because "I've got the Facebook check-in that I use [to] tag people, the people I'm with properly, whereas I can't seem to tag people on Foursquare."

In addition to imagined affordances being "mediated and shaped in relationship to perception, imagination, design, and use" (Nagy and Neff 2015, 6), it is important to keep in mind here, also, that "affordances may be present for only one individual or a group of individuals but not for others" (3). For instance, for Lachlan (media production, aged 25–34), as noted earlier, Facebook was how he kept in touch with friends, Twitter was an information and news source, Instagram served his photographic interests, and Foursquare was a diary of his urban mobility patterns. Meanwhile, for

Mario (aged 35–44), who was working in marketing, "Facebook is sort of family, Instagram is like for sales, Twitter is a lot more about my own work person, for lack of a better term, [. . .] Google+ is certainly, like, personal interest more, you know," while Foursquare served as "a record of where I've been." In this way, both men display a desire to compartmentalize their social media use, albeit according to slightly different categorizations that are similar in some ways and different in others: friends, information, interest, digital record, in the case of Lachlan; family, work, interest, digital record, in the case of Mario.

When speaking to Melburnians about their locative media use, it was very clearly evident that all had a sharp awareness of how individual platforms formed part of a "complex ecology" (McVeigh-Schultz and Baym 2015, 2). This is to say that, in addition to recognizing that individual difference shapes understanding of particular platforms and their affordances, crucially "affordances are not experienced in isolation, but rather in relation to a complex ecology of other tools with other affordances" (McVeigh-Schultz and Baym 2015, 2). This became most apparent when asked about the choices they make between related services, such as between Foursquare and those with similar functionality:

> If we are looking for somewhere to eat, or something like that, Urbanspoon [now Zomato] is better than Foursquare. (Sophia, marketing, aged 25–34)

> [Yelp] kind of combines the best elements of Foursquare and Urbanspoon [Zomato] or whatever other review places there are. [. . .] So, I keep Foursquare for checking in and the game, and I keep Yelp for finding out what's good and reviewing it. (Elen, marketing, aged 25–34)

> [I read Foursquare's] tips and things, sometimes yeah, not often though because you would use Urbanspoon [Zomato] for food reviews. (Allen, architect, aged 55–64)

> So, you could compare [Foursquare] to something like Yelp or stuff like that, which is not as strong over here [in Australia compared with the US], and, for me, Foursquare is probably the better source for tips, more than anything else in Australia. (Mario, marketing, aged 35–44)

From these quotes, we gain a glimpse of the complicated and competitive app ecology that Foursquare is operating within. These end-user perceptions of Foursquare vis-à-vis other services are also particularly interesting in light of the company's more recent attempts, described in Chapter 3, to reinvent what it and its signature app do, with a number of participants clearly favoring other platforms over Foursquare when it comes to food-related search and recommendations—the very thing the company has largely refocused its efforts around.

In this section, I have explored the forms of compartmentalization that inform and are evident in the choices that people make concerning the selection and perceived uses of specific location-sensitive mobile social media platforms. In the following section, I turn to consider the kinds of compartmentalization that are at play when it comes to deciding what, how, and with whom to *share* information (or not) via these same platforms.

COMPARTMENTALIZATION, SHARING, AND THE CHALLENGES OF "CONTEXT COLLAPSE"

"Engagement, and the forms of interactivity that constitute it," Anthony McCosker (2017a, n.p.) writes, "designates the 'publicness' of social media's management of visibility. Visibility is now, arguably, the core value produced through social media sites for individuals." A key aspect of end-user negotiations around the visibility afforded by social media apps to emerge from the interview data concerns location disclosure and the sharing of information that reveals one's precise geographic position. For the Melburnians I spoke with, location was rarely disclosed unthinkingly and without some form of say in what is being disclosed and how widely—as Sophia (marketing, aged 25–34) put it, "If people check me in [on Facebook] it makes me nervous." Rather, participants were, as a rule, quite particular about whether to disclose geolocation information at all, and, if so, in what circumstances, when, and to whom (see Bertel 2016). Given the choice, participants preferred to exert some form of control over—to "curate"—how their location information was disseminated. This is because sharing is "an essential communicative practice," but one "with social consequences" (J. Kennedy 2016, 461; see also Johns 2017). In her article tracing various theorizations of sharing, Jenny Kennedy argues that an understanding of sharing as "intensifying the social"—which she refers to as "affective sharing" (2016, 469)—has been "central to debates on social connectivity" (468). This conceptualization of sharing as social intensification is drawn from the work of Andreas Wittel (2011, 5), who notes that "the decision to share will generally produce an intensification of social activity and social exchange." For Kennedy, a key aim of "affective sharing" is to "provoke" forms of sociability whereby exchange between members of weak and strong tie social networks "intensify social bonds" (2016, 469).

Despite this push for social intensification, it remains the case that social media sharing is a performative *and* necessarily selective process (J. Kennedy 2016). As Alice Marwick and danah boyd (2011, 114) write, "We present ourselves differently based on who we are talking to and where the conversation takes place—social contexts like a job interview, trivia night at a bar, or dinner with a partner differ in their norms and expectations. The same goes for socializing online." Drawing out a similar point, Ari Lampinen et al. (2011, 1) note that "[w]hile social media are all about sharing content with a community, few people wish to share everything with everyone all the time." The difficulty is that "in making decisions about what to disclose and when, individuals often struggle to reconcile opposing goals such as openness and autonomy" (Lampinen et al. 2011, 1). These decisions regarding what information we share and where are made more difficult by technological systems and networked digital structures that "encourage sharing, over withholding, information" (Davis and Jurgenson 2014, 478)—a case in point being Facebook's seemingly endless tinkering with its default settings to encourage "frictionless sharing."

A key impact of this push for "seamless sharing," Marwick and boyd (2011, 115) argue, is that an individual's desire for "variable self-presentation" becomes "complicated by increasingly mainstream social media technologies," like Facebook, Twitter, Instagram, and others, that appear to "*collapse contexts* and bring together commonly distinct audiences" (emphasis added).

One early understanding of this idea of the collapse of contexts was developed by Michael Wesch (2009) in a study of vloggers (video bloggers), where he found "many first-time vloggers perplexed by the webcam" mounted on or embedded near the top of their computers, with these new users "often reporting that they spent several hours transfixed in front of the lens, trying to decide what to say." Wesch (2009, 23) explains the novice vlogger's dilemma in the following terms: "The problem is not lack of context. It is context collapse: an infinite number of contexts collapsing upon one another in that single moment of recording." What this dilemma precipitates among inexperienced vloggers, Wesch (2009, 23) argues, is a "crisis of self-presentation in the face of infinite possible others," and of unanticipated "potential futures, and different contexts" in which their webcam context might be accessed and viewed.

In later social media research, such as Marwick and boyd's (2011) study of young people's use of Twitter, context collapse is cast less as an issue of having "to imagine a virtually infinite number of possible others," as Wesch (2009, 23) puts it. Rather, for Marwick and boyd, the issue is one of how to *manage* a virtually infinite number of possible other social media users. Their argument is that many social network sites, like Facebook and Twitter, complicate opportunities for "variable self-expression" (Marwick and boyd 2011, 115) by collapsing multiple contexts and bringing otherwise distinct audiences together.

Davis and Jurgenson (2014, 480–481) seek to add further nuance to Marwick and boyd's account by differentiating between what they call "context collusions," those situations "when we invite various social contexts to come together," and "context collisions," "when disparate networks overlap" with "potentially chaotic results." Both of these forms of context collapse, "and the management of collapse more generally," they write, "are joint products of architectural affordances, site-specific normative structures, and agentic user practices" (482). From their review, they discover that "while early context collapse literature focused primarily on affordances, and in doing so, on the consequences of network porousness," in more recent work "a *new agentic focus* has emerged" (482; emphasis added).

Much of the focus on user agency in this context has been around how people *manage* context collapse. This can take a multitude of forms of "interpersonal disclosure management" (Lampinen et al. 2011, 3). For instance, Lampinen et al. (2011) list a number of compartmentalization strategies that people employ in dealing with context collapse. These include the following:

(1) "targeting sharing with different users" (Lampinen et al. 2011, 5)—such as by creating and using profiles across multiple services, or what Ben Light (2014, 104) calls the selection of "different sites for different purposes";

(2) "group-specific sharing" (Lampinen et al. 2011, 6)—by dividing one's network and audience into different groups and subgroups, and generating "separate content to different audiences" (Light 2014, 104). In this way, Marwick and boyd (2011, 120) argue, even within a single platform, such as Twitter, users can manage context collapse by writing "different tweets to target different people" within a single information stream; or, as they argue in a separate study, by "hiding content in plain sight"—a practice they refer to "social steganography"—so that their content might be meaningful to some readers of their information stream (such as one's friends) but not to others (such as one's parents) (Marwick and boyd 2014, 1058);

(3) "deciding not to publish" (Lampinen et al. 2011, 6);

(4) "controlling offline behavior" (Lampinen et al. 2011, 7); and

(5) other "corrective strategies" (Lampinen et al. 2011, 7)—such as deleting content posted previously to social media sites.

With the exception of the second in the preceding list, the forms of agency displayed in the management of context collapse detailed by Lampinen et al. (2011) feature strongly in conversations with Melbourne locative social media users. For example, in terms of *targeting sharing with different users*, Caroline (executive assistant, aged 25–34) told of how, when traveling, she would check into airports to tell family that she had arrived safely—common practice for many travelers. This she would do, however, on Facebook as well as on Foursquare, as her parents have access to her Facebook, but not her Foursquare. Caroline also joined social media service Path in order for her and her partner to be able to check in together at venues "because he doesn't use Facebook." In these ways, Caroline has compartmentalized her information through "mode switching" (Park 2017): she shares information with her partner by switching between a number of platforms; and she shares information with her parents by selectively connecting with them, permitting them access to her accounts on certain social network services (such as Facebook) but not on others (such as Foursquare).

In addition, Mario (marketing, aged 35–44) sought to maintain clear boundaries separating family, work, and other networks through his use of specific applications:

> Facebook is certainly personal, in terms of communicating with family, because my family is overseas, and Twitter is certainly more of a work thing. . . . Google+ [which was still operational at the time of the interview] is certainly like personal interest more, you know. (Mario, marketing, aged 35–44)

What Mario is describing here is not just about compartmentalization according to his perception of each app's utility or function. His response also speaks to Kim Barbour's (2015) notion of "registers of performance," where our online personae can shift between various registers, including (among other things) the professional, the personal, and the intimate (58).

Ben Light makes the pertinent point that "crafting messages for multiple participants is complex, requiring attendance to a diversity of audiences" (Light 2014, 104; see also Skovholt and Svennevig 2006); the "visibility labour" (Abidin 2016) involved in maintaining such self-conspicuousness can be quite demanding. As Caroline (executive assistant, aged 25–34) recalls, "You know the time we take to check into a place on Facebook and on Foursquare and then Instagram? When I went to the [Australian rules] football last, it took me to, like, the middle of the first [twenty minute] quarter before I could actually watch the game!"

With respect to *deciding not to publish*, Marcus (production manager, aged 25–34) told of a recent non-business-related trip to Singapore where he managed to secure a business class flight and entry into an airport business lounge ahead of his flight. When checking in on Foursquare, however, he deliberated before deciding to omit mention of the lounge access and flight upgrade "so people don't think, 'oh, you're off to Singapore for work and you're flying business class but you work for a charity.'"

Ben Light (2014) describes decisions, such as Marcus's, to not publish certain content on social media, or to not register one's precise location, as forms of "disconnective practice." For Light (2014, 17), disconnective practice "involves potential modes of disengagement with the connective affordances of SNSs in relationship to a particular site, within a particular site, between and amongst different sites and in relation to the physical world." Each of the Melbourne users interviewed was able to furnish rich examples of his or her disconnective practices, particularly in relation to specific venues that were deemed to be inappropriate for checking into. For example, in Rebecca's (aged 25–34, marketing) case, she states, "I've been into an adult bookstore that I didn't check into intentionally. What else would I not check into? I would not check into somewhere if I were going on a date." What is involved here, Light (2014, 107) argues, are forms of self-censoring "editorial ethics." This is the idea that what we choose to post (or not) to social media sites "can be influenced by what we and others see as a morally right action" (107). To "disconnect" in this way, Light (2014) suggests, is done by individuals in order to protect themselves ("not wanting to be judged for certain activities," 105), and to protect the people they are connected with (108; see also Burchell 2017).

In terms of *controlling offline behaviour*, Harry (student, aged 16–24) remarked that this was a concern for younger adults seeking to establish a career, especially for those hoping to enter public life: "I've got a friend in Brisbane who is in the labour movement and he's worried if he ever sort of goes into electoral politics, people will look at his Twitter account and point out all the stupid things he's said."

And, finally, with respect to *corrective strategies*, Marcus (production manager, aged 25–34) stated his dislike of "the whole mixing of your check-in on Foursquare and [how] it puts it on Facebook or Twitter." He spoke of how, at one point, after having accidentally cross-posted a check-in on Foursquare to other platforms, he "went onto Facebook and Twitter and deleted those Foursquare posts it had done on my behalf or their behalf."

CONCLUSION

In this chapter I have explored how end-users engage with location-sensitive mobile social networking apps. The analysis of this chapter is informed by and draws from interviews I conducted with Melbournians, as part of a larger comparative study, that considered how they used specific apps (like Foursquare) as part of a wider suite of location-sensitive social media services. Thus, a key aim of this chapter has been to examine the "intensity of particularity" (Cooley 2004, 148) associated with user practices involving a specific app, while remembering that usage is always situated within a wider "ecology of communication" (Altheide 1994).

In the inaugural issue of *Social Media + Society* journal, Mary L. Gray (2015, 2) argues that "we need a curatorial theory of social media to shift our attention to the messiness of context and human interaction in the thick of social media's significance." In seeking to make sense of the particularities of locative social media end-user practices, I have framed discussion around the concept of *compartmentalization*. While I don't mean to suggest that compartmentalization constitutes a coherent "curatorial theory" of the sort that Gray is calling for (although it might usefully contribute to one), it is nonetheless productive for thinking through the complicated ways that people draw distinctions between apps based on their own sense of an app's utility or purpose, and the "imagined affordances" of the apps that they use; it is also valuable for making sense of the "social consequences" of sharing, and the desire for "social intensification" without succumbing to "context collapse." In short, the core argument pursued in this chapter has been that, despite the entangled nature of the contemporary smartphone app and social media ecology, end-use routinely involves attempts at compartmentalizing these services, the functions they are seen to fulfill, and the content that we share on them, in ways that work to complicate the apparent "collapse" of social sharing contexts (see also Costa 2018).

In closing this chapter, there is one final point that I wish to make in relation to the preceding analysis of app entanglements. This concerns the way that the types of compartmentalization described in this chapter have been identified by a number of scholars as especially important in the negotiation and performance of identity—an issue that forms the focus of the next chapter. It has been argued that users may "exploit the availability of various interfaces to create partial identities" (van Dijck 2013c, 211), or may create "multiple mini performances of identity" (Papacharissi 2010, 307), and that "keeping up multiple personas across platforms may be a powerful strategy for users to 'perform' their identity in a Goffmanesque manner" (van Dijck 2013c, 211). At the same time, there is increased concern for how social media platform architectures both "enable and constrain the sculpting of personal and professional persona[e]" (van Dijck 2013c, 200). In terms of the latter, considerable attention has been given to platform owners' attempts to constrain identity play. For instance, as José van Dijck writes:

> Platform owners are keen to commit users to present uniform personas instead of splitting up their online identities through various platforms, which messes up the clarity and coherence of their data. (van Dijck 2013c, 212)

This apparent push toward uniformity of identity (and data purity) is most evident in the pursuit of real name policies by the likes of Facebook (van Dijck 2013c, 212) and Google (van der Nagel 2017), and through "federated identity management" systems—that is, "the interlinkage of databases and authentication procedures across and within domains" (van Zoonen 2013, 47)—such as Facebook's "single sign-on" (Wilken 2014a). However, what results from such efforts, it has been suggested, is a "fundamental misjudgment of people's everyday behaviour" (van Dijck 2013c, 212), and an undermining of "the understanding of identity as multiple" (van Zoonen 2013, 47).

It has subsequently been pointed out that these critical assessments tend to neglect a further, important countervailing point: that compartmentalization and forms of end-user segmentation are not undesirable to platform owners. "One element often missing in analyses of datafication, and postdemographic personalization, or 'profilization,'" Anthony McCosker (2017a) writes, "is the indeterminacy and unfinished character of segmentarity" or compartmentalization:

> In a postdemographic world, characterized by the continuous production and calculation of social data as likes, interests, keywords, locations or hashtags, social media platforms are *designed* with techniques of market segmentation in mind. (McCosker 2017a, n.p.; emphasis added)

The result, then, McCosker (2017a) argues, is a "tension or paradox" between, on the one hand, "the personal, curative or performative uses of social media," and, on the other hand, "the design and commercial usefulness of platforms and apps." These are not mutually exclusive or opposing forces or interests. Indeed, in key respects, "multiple mini performances of identity" (Papacharissi 2010, 307) are of specific value to platform owners and marketers:

Social media analytics are first and foremost techniques of market segmentation, so these tools and the platform features they leverage work to multiply, quantify and in turn deepen processes of segmentation by incremental adjustments to interface design, affordances and subsequent cultures of use. (McCosker 2017a, n.p.)

Thus, "social media performance takes place in relation to continuously segmenting forces and practices," and, yet—and this is the crucial point— "this does not necessarily reduce the utility of social media platforms" (McCosker 2017a). Thus the commercial imperatives of social media platforms do little to diminish the ongoing pursuit and appeal of end-user "identity curation"—it is in the platform owners' interests that this is permitted to continue. In the next chapter, I take up this issue further by turning my attention away from app entanglements and toward a closer examination of the careful crafting of venue check-ins and other means of registering location to "shape others' perceptions of who [one is] and where they go" (Schwartz and Halegoua 2015, 1647) in order to create "idealized performances" of the "spatial self." While this chapter has examined the concurrent use of multiple location-enabled smartphone applications, the following chapter takes a closer focus, to examine diverse end-user practices as they relate to a single locative media application—Foursquare—as they occur at specific points in space and time.

CHAPTER 6

Territories of the City and the Self

Locative Mobile Social Networking, Urban Exploration,
and Identity Performance

Location-aware media and mobile devices can be construed as interfaces, or filters, which co-constitute technology, social interaction, location, and everyday spatial practice.
—Hulsey (2015, 156)

Location has changed from being something you *have* (a property or state) to something you *do* (an action).
—Cramer, Rost, and Holmquist (2011, 65)

Location awareness embedded in mobile devices strengthens the connection to physical spaces, creating new geographies of mobility.
—Pertierra (2012, 109)

Location has become an important piece of personal and spatial identity construction.
—de Souza e Silva and Frith (2012, 163)

INTRODUCTION

Foursquare, the one-time location-sensitive mobile social networking and now search and recommendation service, has played a formative role in promoting the wider take-up of locative media, and in shaping everyday navigation and engagement with urban space and place. Foursquare, in short, has come to be regarded as a complicated site in which territories of the city and the self are negotiated and performed. In this chapter, I examine how this has come to be the case.

The analysis of the chapter is presented in two sections. In the first section, I examine how urban spaces and places are explored, catalogued, and communicated—that is, "how people transmit location" (Frith 2014, 894) through Foursquare. In the second section, the focus turns to a consideration of how these communicative practices are entwined with individual identity construction and performance. I conclude by responding to some criticisms of Foursquare and location-based mobile social networking.

While app development and use are clearly co-implicated, for the sake of clarity, I have separated the business side of Foursquare from end-user engagement with Foursquare. Thus, whereas Chapter 3 examined the evolution of the Foursquare business and the search for a stable revenue generation model, this chapter considers consumer end-use of this service. In examining end-use, I draw, once more, on my own interview-based study of Melbourne Foursquare users (part of a larger comparative study that is described in the Introduction to Part II), as well as on the now extensive existing research literature on Foursquare end-use. All of the studies of Foursquare that inform this chapter—my own, and those of others—were conducted prior to Foursquare Labs's 2014 "unbundling" of its public-facing services into two interconnected apps: a rebranded Foursquare (for local search, discovery, and recommendations) and Swarm (for social networking and check-ins) (see Chapter 3 for more detail).

Given the specific temporal positioning of much of this scholarship, this chapter ought to be approached as a work of contemporary new media history. And while it is acknowledged that much has changed with respect to location services between 2014 and the completion of this book, Foursquare end-user practices continue to warrant careful attention for what these practices reveal about the emergence of locative media among a wider smartphone-using public, and, more specifically, for what they reveal about the complicated identity work, richness, variety, and contradictions associated with "vernacular data cultures" (Albury et al. 2017)—issues of ongoing concern within mobile communication and internet scholarship.

COORDINATING, CATALOGUING, CONNECTING: WHAT PEOPLE DO ON FOURSQUARE

Simon Dell (2011, 31) suggests that "it's part of our human nature to create a pattern of acknowledging our whereabouts." In their study of Foursquare, Janne Lindqvist and colleagues (2011) uncovered nine different motivations for why people tended to "check in" to a venue using the application. Indeed, across the available critical literature on Foursquare, many and varied consumer end-uses of Foursquare have been identified (in addition to Lindqvist et al. 2011, see also Cramer, Rost, and Holmquist 2011; Patil et al. 2012). In this chapter, and in order to simplify things, I condense these various recorded motivations into three broad categories that I have adapted from Lee Humphreys' (2012) study of communicative practices across mobile social networks. These categories relate to *coordinating* practices, *cataloguing* practices, and *connection* practices. These three categories are valuable for capturing economically the diverse array of things that people do on and through Foursquare. Each of these three categories will be discussed in turn.

Coordinating

Humphreys' use of the term "coordination" draws from the earlier, influential work of Rich Ling on the use of mobile phones in the "microcoordination" of everyday activities and, in particular, of basic daily travel arrangements (see Ling and Haddon 2003; Ling and Yttri 2002). Thus, as Humphreys (2012, 503) intends the term, coordination suggests "communicative exchanges around organizing and situating our physical selves in relation to one another."

In Humphreys' (2007, 2010) seminal study of Foursquare's precursor, Dodgeball, "the coordination function" was considered crucial to how this SMS-driven pre-smartphone service was used. In subsequent studies of Foursquare, however, it has been found that "coordinating congregation," as Humphreys puts it (2010, 769), is far less common, due, at least in part, to the complicated social codes associated with Foursquare "check-ins." For instance, check-ins can enact "the performance of a particular kind of sociality" (Licoppe 2014, 118). What this "sociality" is, however, is open to negotiation, with the check-in described "as a type of written event, akin to a speech act" (Licoppe 2014, 111), or "language game" (112), one regarded by those reading the check-in "as a kind of social action to be (possibly) responded to [or not] and not just as a factual claim about one's location" (111). That is to say that the check-in "often does not work as a stand-alone piece of information" (Frith 2014, 899), such as to say, "I am here; you can meet me." Rather, it serves as "a form of mobile communication that invites further mobile communication" (898) prior to any (generally rare) opportunity to meet up.

While "coordinating congregation" (Humphreys 2010, 769) is not a common feature of Foursquare use, adopting a more expansive understanding of "coordination" reveals a more fundamental aspect of all locative media use: the way that these media facilitate—that is to say, *coordinate*—our movements through urban space. Locative media do this in a very literal sense, by determining our position in time and space through the fixing of longitude and latitude coordinates. Equally importantly, though, they also achieve this via other means: by fostering urban exploration and place discovery (Frith 2014, 901). Not only have "locative media made the process of navigating everyday space [. . .] a seamless, day-to-day activity" (Farman 2012, 87), such as through the provision of mobile maps, they also "offer alternative ways of experiencing the city" (Özkul 2015a, 113). Catering to these exploratory desires are a number of smartphone apps that encourage urban discovery (Frith and Kalin 2016, 47–48) and generate "a new sense of places [and] a new way of place-ing oneself" (Özkul 2015a, 107).

This process of urban exploration and discovery is a recurrent theme in participant accounts of Foursquare end-use, and was made explicit in my own participant interviews with Melbourne users. For instance, in one memorable interview moment, Sophia (marketing, aged 25–34) reflected on her use of the app, declaring, "I am Captain James Cook!" The implication here is that Foursquare was the vessel supporting her own urban explorations, much like Cook, in established narratives of Australian colonial settlement, was an explorer captaining *HMS Endeavour* in search of *Terra Australis Incognita*—the "unknown southern land" (Edmond 1997). Urban discovery has also figured prominently in studies of UK users (L. Evans and Saker 2017, 81) and was a significant factor in surveys of usage patterns among US users (Lindqvist et al. 2011).

Foursquare has, across the various iterations of its apps, sought to actively encourage exploration, including through the careful spacing and curation of badges that require travel if they are to be obtained (see Frith 2013), as well as via the introduction of the "Explore" search feature, saved lists of possible venues to visit in the future, limiting the points earned for already visited check-ins, and individual end-user "taste" or "expertise" profiles that are built up through check-ins and the leaving of recorded tips and recommendations. Check-ins at known venues and newly discovered venues combine to "construct a trajectory" or mobility paths, flows of movement, punctuated by check-in "pauses," that actively create links (data trails) between these various locations (Frith and Kalin 2016, 51). This process of checking in "thus becomes an integral resource in reconsidering mobility as something that occurs before, during, and after actual movement"

(51). For Elen, this took the form of a checklist of Foursquare badges to collect, with a plan for how to acquire as many as possible:

> I have a list in the back of my planner of all the badges, so it's like collecting a set of something and then I have as many as possible. [. . .] I've also got a strategy. It's basically just a way of, I suppose—as an extra bonus to my general day-to-day activities, planning around how I can possibly sneak in a badge. (Elen, marketing, aged 25–34)

This form of urban exploration has been dubbed "chance orchestration" (van der Akker 2015); it is a term that captures a dual process. On the one hand, there are the ways that Foursquare seeks to encourage—to "orchestrate"—urban exploration through the design of its interface and software, through algorithmic sorting (van der Akker 2015, 35), and through the creation of badges (or stickers in Swarm app), some of which are more difficult to get than others. On the other hand, there are the ways that Foursquare users "become aware of," or chance "and act upon, the digital layers [populated] with folk knowledge" (35), which are presented to them through the app. Through the preceding discussion, I have explored a variety of ways that people have used Foursquare, and which Foursquare Labs has sought to encourage, in the coordinating of congregation and the coordinating of our movement in time through space. I now turn to the second category of Foursquare use: cataloguing.

Cataloguing

"Cataloguing," in the generally understood sense of the term, involves the recording and classifying of information "according to a categorical system" (Humphreys 2012, 504). This is something that is undertaken both by platform operators and by users of these networked services (504). For Foursquare Labs, for example, its catalogue of socio-spatial information—and the data analytics offerings built atop this catalogue—are the fruit of two combined datasets: its places graph and its social graph (see Chapter 3; see also Shaw 2012). For end-users of Foursquare, prior to its redesign as a search and recommendation service (and for users of Swarm subsequent to this redesign), cataloguing occurs principally through venue check-ins. In the scholarship on Foursquare end-user engagement, the venue check-in holds particular importance as "a sociotechnical practice of place-based digital memory" (Frith and Kalin 2016, 51). Check-ins thus serve as vital "archives of the everyday" (46). As such, it is not just the *persistence* of

Foursquare check-in data that is significant, but also the *searchability* of this data, especially in relation to mobility patterns, places visited, and the memories associated with both (Frith and Kalin 2016, 49; Humphreys 2012, 505).

These characteristics—persistence and searchability—hold obvious appeal to Foursquare. Through the incorporation of "additional mnemonic metadata" and "by encouraging users to remember previous check-ins, [Foursquare] produces a *digital network memory* that connects the past to the present" (Frith and Kalin 2016, 50, emphasis added), thus building rich historical data that feed into its real-time analyses of place-based social interactions, and of trends and "off-trends" (anomalous events) (Sklar, Shaw, and Hogue 2012).

Yet, as research on Foursquare end-use has shown, while the persistent, searchable nature of the service is certainly used to scope prospective venues, a recurrent finding is that check-ins also serve an important archival function for the recording of "ordinary affects" (Stewart 2007). For instance, Melbourne Foursquare user Elen noted the usefulness of the Foursquare archive given her own experiences with attention deficit disorder (ADD):

> I've got ADD, so I like having lists and knowing that I can check things and so that's why I like having a list of where I've been [. . .] because I don't remember a lot. (Elen, marketing, aged 25–34)

Meanwhile, Mario (market research, aged 35–44) stated that Foursquare was "not about a network connection as such, but more about a record of where I've been, maybe where I'd like to go." This is consistent with the findings of other studies. For instance, Didem Özkul and Lee Humphreys' study of mobile media and memory found that "the ability to share or transmit [. . .] memories via the mobile was less important than the documenting and memory-making nature of the devices" (Özkul and Humphreys 2015, 356; see also Cramer, Rost, and Holmquist 2011; Saker and Evans 2016). Özkul and Humphreys also make mention of one London-based mobile user who described the taking and posting of her photos as "a sort of visual diary" composed of "random things" (2015, 355), and another who disclosed that he sometimes posted semi-abstract photos—"just a little snippet of a tree or a park full of leaves"—without any explanatory text as the best way of summing up the places he'd been and that were meaningful to him (358). In these examples, Foursquare functions as a "mediated memory object" (L. Evans and Saker 2017, 50): it permits a "curated" history of "personal movements" (47) and personal moments.

In addition to Foursquare check-ins serving as a record of our daily interactions, they also create a rich "documentation of our relationship with the places with which we have special ties" (Schwartz 2015, 96). Raz Schwartz refers to these relationships (after Low and Altman 1992) as "online place attachment" (Schwartz 2015, 95). These forms of connection are reinforced in a variety of ways, including symbolically, through the "mayorship" function of Foursquare (and now Swarm) (Özkul 2015a, 108), and socially—thus becoming "collective attachments" (Schwartz 2015, 96)—through the automatic sharing to followers within one's social network.

The lengths some Foursquare users would go to in order to register and reconfirm these "online place attachments" were quite extraordinary. For instance, Allen (architect, aged 55–64) would take a specific walking route to work each day that would enable him to "check in" at a specific public building that he had a particular fondness for. While living in Brisbane, Harry (student, aged 16–24) spoke of diverting his route whenever possible in order to check in at a particular piece of public sculpture associated with the Queensland Museum. Harry also spoke of his determination to check in at a Chinese airport, despite being unable to do so while physically at the airport (due to Foursquare being blocked at the time by China's "Great Firewall"). Once he arrived home he "opened up Firefox, went to the mobile Foursquare site, masked [. . .] geo-location and checked in" to create a remote ex post facto record of his fleeting visit to China. Meanwhile, Rebecca detailed the elaborate steps she would take each day while living in San Francisco that would permit her to check in to the Golden Gate Bridge while driving across it:

> I had my whole process set up. As I was leaving my apartment, before I would drive away I would get Foursquare queued up, and I would have "Golden Gate Bridge" typed in to the search field, so that, when I was close, I could just press 'search' and then check-in, so it would be nice. (Rebecca, marketing, aged 25–34)

The preceding snapshots provide a small glimpse of the complicated forms of end-user engagement with Foursquare that feed the firm's data-gathering ambitions while simultaneously extending beyond these to create personal and collective meaning for users of the service (see also Frith and Saker 2017). Drawing on Gaston Bachelard's (1994, 8) concept of "topoanalysis"—which he defined as the "systematic psychological study of the sites of our intimate lives"—Özkul and Humphreys (2015, 362) thus regard location-sensitive social media use as a productive "lens through which to understand things about ourselves that we may not even [yet] be aware of." It is a sentiment that connects with the focus on identity in the second half of this chapter. Before turning to these concerns, I address the last of the categories for sorting recorded motives for Foursquare end-use: connecting practices.

Connecting

So "much of public sociality through mobile social networks is about connecting with others" (Humphreys 2012, 502). Informed by urban sociology, this understanding of the term "connecting" relates to "managing one's social distance with others" (502). Connecting "allows users to decrease and occasionally increase social distance between themselves and others" (502). With respect to *increasing social distance*, Humphreys' (2010) study of Dodgeball reports a number of instances where participants used the service to either avoid a particular person or to "eliminate potential places to socialize" (774) based on venue check-in information they had received. With respect to the more prevalent understanding of connecting as *decreasing social distance*, examples abound. These range from responding to the check-ins of others within one's social network, performing a kind of civic duty within the Foursquare community by leaving likes, tips, and recommendations for other users, to using location-enabled smartphone apps to express one's care for loved ones or enabling friends to "stay connected with each other whether or not they are living in the same city" (Özkul 2015a, 111). There are also more overt, declarative modes of communication that promote "display" (Özkul 2015a, 109) and "showing off" (Humphreys 2007), as illustrated in the following reflections by Melbourne Foursquare users on their use of the app while traveling overseas:

> To be able to check into Kihei Beach on Maui was just like, "be jealous." (Rebecca, marketing, aged 25–34)
>
> When I was traveling, I like[d] to check in to other places. I wanted people to know that I'm a traveler. [. . .] So, yeah, I kind of want people to know, "Oh, yeah, she's just arrived in LAX and now is in Mexico, what a lucky [person]," [and] make my friends jealous. (Caroline, executive assistant, aged 25–34)

Foursquare use has also been credited with reducing social distance between relative or complete strangers, specifically in ways that lead to increased place attachment. For example, Schwartz found that 20 out of the 25 US informants he interviewed "confirmed that after they check in to a place they use the application to look up the mayor [of that venue]" (Schwartz 2013, 143). In trying to make critical sense of this finding, Schwartz draws on social psychologist Stanley Milgram's concept of the "familiar stranger" (see Milgram 1992), which describes a relationship-at-a-distance that forms between two strangers who recognize each other through their daily encounters but choose not to interact (Schwartz 2013, 139–140). Schwartz's Foursquare research prompted him to update this concept to that of the "networked familiar stranger," to capture how contemporary interactions with proximate strangers are increasingly mediated through services such as Foursquare. As one of his New York City interviewees reports,

> The mayor of my favorite coffee place is this guy I've seen there several times and have been keeping an eye on for some time now. It all started when I saw he was the mayor of my local sushi place and since then I noticed that he is also the mayor of several other places I really like. [. . .] We've never talked and I think we never will but he is dreamy. (Sara, quoted in Schwartz 2013, 143)

In developing this concept of the "networked familiar stranger," Schwartz reiterates two key points made by Milgram. The first being that "the status of familiar stranger does not reflect an absence of a relationship," but, rather, "a different form of interpersonal relationship" with "properties and consequences of its own" (Schwartz 2013, 140). The second is that "familiar strangers are as important to the perception of the local surroundings as street signs, public parks benches and other local landmarks," and that "all of these in turn support the feeling of local identity and belonging" (140). In this way, the "networked familiar strangers" encountered via services like Foursquare can function, alongside check-ins and mayorships, as important means of building a sense of local place attachment (Özkul 2015a; Schwartz 2015).

A similar line of argument is supported by interviews with US-based Foursquare mayors. Humphreys and Liao (2013) found that Foursquare use fostered various forms of "parochialization." This they define as the "process by which people share socio-locational information with one another through communication technologies such as check-ins on mobile social networks, such that the public realm, where people had previously encountered strangers, starts to feel more familiar due to the social exchanges through the network" (Humphreys and Liao 2013).

Both of the preceding examples—that furnished by Schwartz (2013) and that by Humphreys and Liao (2013)—draw out the possibility of forms of connection (the last of the three categories into which motives for Foursquare use have been grouped). These connecting practices work to close the social distance between relative or complete strangers (and not just between those known to each other), at the same time further strengthening place attachments.

The chapter discussion to this point has been concerned with an extended examination of how locations have "identities of their own" (de Souza e Silva and Frith 2012, 167) that are formed through a variety of processes, including our day-to-day individual and communal negotiations with them and as a result of the "digital information embedded in them" (169). In the second part of this chapter, consideration is given to how our mediated engagements with locations (especially through use of locative media) have become "important aspects of people's identit[ies]" (167). A key aim of the following analysis is to explore how the practices discussed earlier—coordinating, cataloguing, and connecting—are all enrolled in individual end-user attempts to "construct their ongoing present sense of identity" (Frith and Kalin 2016, 44).

IDENTITY PERFORMANCE, LOCATIVE MEDIA, AND THE SPATIAL SELF

Self-identity is now widely understood not as something stable, fixed, and singular, but, rather, as fluid, performative, and multiple. This understanding is fed from a range of sources, including (but not limited to) the following: (a) Erving Goffman's (1971 [1959]) seminal dramaturgical account of self-presentation and impression management; (b) feminist theory, particularly the work of Judith Butler, who asserts that gender reality is performative, rather than "expressive of a gender core or identity" (Butler 1988, 527; see also Butler 1990), and whose thesis has at its heart the claim that "every performance of identity is also always potentially disrupting or disturbing" (Elliott 2014, 127); and (c) influential accounts within social theory that have sought to understand how "we make a meaningful self-identity in a late modern, postindustrial world" (de Solier 2013, 14) where, increasingly, at a broader societal level, "the very idea of controllability, certainty or security" is being challenged (Beck 1999, 2; Beck and Beck-Gernsheim 2002). With respect to the last of these, Ulrich Beck suggests that, faced with such upheavals, the individual in late modern times must become an "actor, designer, juggler and stage director of his or her own biography, identity, social networks, commitments and convictions" (Beck 1994, 14). One's biography—how one sees oneself and how one presents oneself—becomes, for Beck, a chosen biography, a "do-it-yourself biography," or what Anthony Giddens (1991) calls a "reflexive" biography: one that is self-aware, self-consciously made. The suggestion here is that, by force of circumstance, individuals must "choose, construct, interpret, negotiate, display who they are to be seen as" using a wide array of resources, both material and symbolic (Slater 1997, 84).

The clear connection that is drawn here between material culture and identity construction and performance has been of significant interest to researchers of media and communication technologies. Prior to the arrival of smartphones, a number of studies of mobile media and communication have examined in detail the close connections between our use of mobile devices and self-identity formation. For example, in their cross-cultural study of mobile phone use in the United States and Japan, James Katz and Satomi Sugiyama (2005) argue that the mobile phone is strongly connected with status and identity (63), and that each phone is "not merely a tool but as well a miniature aesthetic statement about its owner" (64; see also Hjorth 2009). Studies of how mobile phone users negotiate public/private tensions also suggest that "the difference between when we are acting and we are being ourselves is on the whole less distinct" because the mobile phone provides us with the means to "'stage' ourselves" (Fortunati 2005, 217). And, in *Mobile Interface Theory*, Jason Farman (2012) adopts a deliberately broad definition of mobile media as comprising a whole raft of everyday objects and technologies that "signify elements of our identity" (1). Building from this definition, Farman develops a theoretically nuanced account of our embodied engagements with mobile media and various "modes of inscription" (2012, 30) that implicate each of us as individual and social subjects. These are just a few select examples drawn from a rich vein of mobiles scholarship that recognize, explore, and document the close interconnections between mobile communication use and identity work, and the complicated ways that mobile phones are "closely involved in self-identity construction" (Castells et al. 2007, 112).

Practices of self-presentation, impression management, and identity performance have also come to form dominant themes within research on social networking sites. Within the literature on social media, Goffman's theories have proven especially influential (Hogan 2010). To cite two brief examples, Zizi Papacharissi (2010) notes how social networking services (SNSs) constitute important "sites of self presentation and identity negotiation" (304). Through access to services like Facebook, individuals are provided with a variety of tools—"text, photographs, and other multimedia capabilities" (304)—that facilitate and permit "more controlled and more imaginative performances of identity online" (307); that is, these sites "expand the *expressive equipment* at hand" for the construction and negotiation of identity (307). In a related vein, José van Dijck (2013c, 200) also notes the importance of social media platforms for "the public staging of one's identity" (200); however, she argues that, rather than just serving as "neutral stages of self-performance," social media in fact constitute "the very tools for shaping identities" (213). What unifies much of this work is the premise that, through the entire gamut of social media mechanisms (images, video, status updates, profiles, friend lists, etc.), "social media participants present a highly curated version of themselves" (Schwartz and Halegoua 2015, 1645).

There is also an emerging scholarship that combines interest in these issues with an explicit concern for questions of *location* to explore the "harnessing of place" via mobile locative media "to perform identity to a social network" (Schwartz and Halegoua 2015, 1645). The importance of the ties binding place, communication, and identity are, of course, well established (see, for example, Moores 2012; Wilken and Goggin 2012a, 2012b). Nevertheless, the advent of location-sensitive mobile social networking has provided a key mechanism for the further strengthening of these connections and for the construction and performance of identity.

The argument that I wish to develop in the second half of this chapter is that these issues of identity construction and performance have long sat at the heart of end-user engagement with Foursquare (prior to its unbundling) and other forms of location-sensitive mobile social networking. And while mobile location data capture is now largely automated, and rarely reliant on active check-ins, these processes of identity construction and performance are part and parcel of the contemporary settlement of mobile locative media, and understanding them is crucial if we are to more fully grasp the meanings that people have developed through, and have ascribed to, these pioneering services.

The conceptual anchor for this discussion will be Raz Schwartz and Germaine Halegoua's (2015) notion of the "spatial self"—a notion that is informed by the theoretical trajectories described earlier (especially Goffman and Butler), as well as earlier locative media scholarship on the "presentation of place" (Sutko and de Souza e Silva 2011) and the "presentation of location" (de Souza e Silva and Frith 2012). The "spatial self" is a term that Schwartz and Halegoua (2015) coin for *categorizing* what they regard as common cultural practices, and as a *critical lens* through which to study "myriad expressions and performances of identity and place online via social media" (1647) and "the growing number of geocoded representations on locative and social media" (1653). The spatial self, Schwartz and Halegoua explain, refers to a variety of instances "where individuals document, archive, and display their experience and/or mobility within space and place in order to represent or perform aspects of their identity to others" (1644).

Staying true to the single-platform focus of this chapter, here I examine the performance of self-identity in relation to Foursquare use, drawing predominantly (once more) from my Melbourne participant data. In exploring these issues, particular emphasis is given here to three key aspects of the "spatial self" concept: how negotiations of the self tend to be *"highly curated"* (Schwartz and Halegoua 2015, 1654); how they often involve *idealized performances* of who a user is or wishes to be based on where they go (1647); and, how, as with "other aspects of our identities, the spatial self is not a unique, singular representation but rather a *multifaceted and fragmented depiction of the self* that has many different versions" (1649).

The first thing to note here is that one's negotiation of one's "spatial self" involves *"intentional* socio-cultural practices of self-presentation" (1647, emphasis added). For Harry (student, aged 16–24), the very fact of having to open the Foursquare application in order to check in makes this process a very purposeful practice: "whereas we always have relation to space just because we occupy that space, what Foursquare and location services allows is for it to be very obviously performed." This, he argues, is because, by registering our coordinates through check-ins, what we are *really* registering are particular activities and actions that we are conveying to an audience. By checking in, what he is reporting on are activities and actions ("I'm at this place doing something"), and this information, he believes, conveys "a more substantial notion of who I am" than simply registering longitude-latitude coordinates.

Other Foursquare users also noted the impact that check-ins have for the presentation of self. For example, Lachlan (media production, aged 25–34) offered the following candid assessment of the self-identity work that he sees taking place through social networking services like Foursquare/Swarm:

> I think, especially these days, people are really conscious of who they are and how they want to be seen to everyone else because everyone is so connected now that I think it's a big part of trying to represent who you think you are to an audience, and whether or not it's false or it's amplified is another story.

Meanwhile, for Marcus (media production, aged 25–34), identity work is clearly performed, yet somewhat less readily admitted to, at least not publicly. Asked why he checks in on Foursquare at the football match, Marcus responded:

> I suppose I want people to know where I am and what I'm doing, and it's to show me advertising my support for the [Australian rules football] team [I follow],

and that I'm there and I'm at the game. And, maybe, also, it's hard to admit, but showing off, that, "hey, I'm out and about and being social, I'm not the guy who's just sitting at home all day watching TV." So, it's more, like, "look where I am today," and then next week, "look where I am, I'm somewhere else," and then the following week somewhere else again. Unconsciously that's what it is, even though I probably wouldn't admit that anywhere else. I'd never admit this outside this room actually.

Such acute awareness of the self-identity work that is being performed through locative social media use is often manifested in strongly curatorial behavior: that is, in the careful presentation of "a certain self to others," as well as in the modification of spatial practice to "project the self" one wishes to identify with, both for oneself and for others (Saker 2017, 8). This may mean highlighting particular activities that are deemed to have a certain social currency (such as being at a new bar or a fashionable restaurant) (see K. Burns 2013). It equally extends to what Ben Light (2014), as discussed in the previous chapter, terms "geographies of disconnection"—those places, situations, and times where individuals do not wish to record and share their presence via social media (see also Burchell 2017). A prime example of this is found in Elen's (marketing, aged 25–34) admission that she would never check into a McDonald's "because it's really unhealthy and . . . it's gross, but when I'm really hung-over I do [eat there], so I don't want anyone to know." At the same time, and somewhat paradoxically, the presentation of a "certain self to others" may also involve the decision to *not* differentiate between venues. For instance, Rebecca reports that she has made a conscious decision to, as much as possible, *not* filter her check-ins, choosing instead to register her presence everywhere she goes:

> Who I am is in part formed by what I do. [. . .] I'm checking into everywhere I go instead of checking into places where I want to set a perception of who I am. It almost is like open exposure and, like, these are the things I do: the good, the bad, the indifferent, the boring, the exciting, the whatever. (Rebecca, marketing, aged 25–34)

Adopting such a stance of "open exposure" is very much still consistent with Schwartz and Halegoua's (2015, 1648) understanding of the "spatial self" as a "stylized repetition of presenting certain places, with certain connotations and meanings, as constitutive of one's identity performance."

The case of Rebecca also draws out convincingly the idea that end-use of services like Foursquare often take the form of "multifaceted and fragmented depiction[s] of the self that ha[ve] many different versions, each with its own characteristics and audience" (Schwartz and Halegoua 2015, 1649; see also van Dijck 2013c, 211). For Rebecca, the aim, it would seem, was to present her "spatial self" as unapologetically multifaceted and eclectic. Whereas Elen (marketing, aged 25–34) suggests that the earlier bundling of various services and functionalities that characterized the Foursquare app prior to its later "unbundling" and splitting into two separate apps (Foursquare and Swarm) actually aided the performative identity work undertaken by its users. Elen states:

> [Foursquare] really does speak to someone's personality: if they care about points, or if they care about offers, or if they care about people seeing where they are, or if they're introverted or extroverted, or if they're more an expressionist type. [. . .] How you use apps like Foursquare really does speak to your personality and your learning behaviors. (Elen, marketing, aged 25–34)

Meanwhile, for others, such as Sophia (marketing, aged 25–34), whose work brought her considerable (and at times unwanted) public exposure, the multifaceted depiction of the self was carefully managed as a protective measure. Rehearsing the compartmentalization arguments of the preceding chapter, Sophia spoke clearly of the need to carefully delineate what she perceived as her "public" from her "private" audiences:

> Foursquare forms part of my personal identity and my public identity which is kind of exceptional for the social media forms [that I use] because I don't put personal stuff on Twitter [or other social networks]. [. . .] On Foursquare, those lucky 16 people [on my follower list] get to see a significantly larger amount of what I check into. [. . .] I don't broadcast as much and they get to see that. (Sophia, marketing, aged 25–34)

This passage, and others quoted above, serve to illustrate just some of the manifold and important ways that end-users of mobile social networking services like Foursquare represent their place-interactions and how these place-interactions are co-implicated in, and co-constitutive of, one's identity performances.

To this point I have examined how urban spaces and places are explored, catalogued, and communicated—that is, "how people transmit location" (Frith 2014, 894) through Foursquare—and how these communicative practices are entwined with individual identity construction and performance. Foursquare has not, however, been without criticism. I therefore close this chapter by detailing the most significant of these criticisms before offering a three-part defense of mobile, location-based social networking services and why, despite significant shifts over the past decade, they continue to be significant.

IN DEFENSE OF MOBILE LOCATION-BASED SOCIAL NETWORKING SERVICES

A key concern expressed in relation to locative media is the commodification that structures use of these services, where value is said to be found "only in the accumulation of a user's movements, locations, and habits" (Farman 2012, 61). This has been a particularly strong criticism of Foursquare. For example, it has been claimed that, within Foursquare's (and now Swarm's) points system, there is an inherent tension whereby "Foursquare mediates the exchange of social for economic capital" (Phillips 2011, 180). In a damning assessment, David Phillips refers to this as Foursquare's encouragement to "a sort of competitive sedentary egocentrism" (2011, 180). "One's economic capital may increase," he writes, "if you get free drinks for enticing friends to buy." Thus, Phillips concludes, through its incentive structure, "the places [Foursquare] makes most visible are places of consumption," and, because of this, it "re-entrenches the hegemonic relation of work and leisure, production and consumption" (180). Reinforcing this critique is the suggestion that the way Foursquare "is linked to potential commercial gain" fixes the overall database of potential places that might be accessed "within a rigid structure of the service industry" (Gazzard 2011, 410), thereby undermining Foursquare's credentials as an urban exploration mobile gaming application (see also Leorke 2015). The general lamentation expressed by these critics, it would seem, is that Foursquare functions, to borrow Tania Lewis's words, as little more than a "banal instrument of a globalizing neoliberal commercial culture" (T. Lewis 2008, 153).

However, this line of critique is limited for at least three reasons. First, such accounts do not do justice to the complicated identity work and richness and many contradictions associated with "vernacular data cultures" (Albury et al. 2017). Sharing one's location is not just about recording one's presence at a commercial venue, it "may also signal mood, lifestyle, or life events and maintain or support intimate social relationships" (Schwartz and Haleqoua 2015, 1646). As I have sought to demonstrate in this and the previous chapter, users of mobile social networking services do not simply acquiesce to the commercial imperatives of platform owners. Rather, end-users treat these services, by necessity, as "sites of improvisation" (Pink et al. 2017, 4) and negotiation involving repertoires of "distinct user tactics"

(Vainikka, Noppari, and Seppänen 2017, 123). Tactical responses can take a variety of forms. For instance, one of my participants, Harry (student, aged 16–24), made a concerted effort to try "to get some distance from the way in which people [generally] use Foursquare" by checking in to non-commercial venues and at events, such as political rallies. Meanwhile, other end-users reported on elsewhere have sought to circumvent the recording of location through the use of third-party GPS spoofing apps (He, Liu, and Ren 2011; Ren 2011; Restuccia et al. 2016). And, in a study of Indonesian Foursquare users, a key tactic was to "game" the system through "jumping" (the process of unlocking symbolic achievements that contravene Foursquare rules for acquiring badges) (Halegoua et al. 2016). While tactical endeavors such as these have been referred to as "misbehaviors" (Papalexakis, Pelechrinis, and Faloutsos 2014), Halegoua et al. (2016, 10) make the important point that, for Indonesian users, "jumping" emerged from a perceived sense of "marginalization within the platform" as they resided outside the United States, the country with the highest clustering of Foursquare badges. For this reason, they argue, jumping "should be recognized and considered [. . .] not as routine cheating but as indications of the platform's politics" (10).

Second, the aforementioned critiques of Foursquare's commercial logics do not adequately account for the symbiosis that is said to exist between identity construction and consumption. This is a relationship that has been articulated most clearly by lifestyle consumption theorists. For example, Adam Arvidsson (2006) notes how "consumers use goods productively" in order "to construct social relations, shared emotions, [and] personal identity." Don Slater (citing Alan Warde) goes further than this, arguing that "all acts of purchase or consumption [. . .] 'are decisions not only about how to act but who to be'" (Slater 1997, 85; see Warde 1994, 81).

Foursquare use is very much enrolled in these struggles over "how to act" and "who to be"—struggles that are often played out around questions of experiences of authenticity. In Michael Saker's study of UK Foursquare use, for instance, access to place-based tips and recommendations that led users off well-trodden paths and away from obvious venues (like Starbucks) created a belief in those acting on this information that they were tapping into local knowledge that opened up experiences of a city "seemingly hidden" from those who didn't use the app (Saker 2017, 943). By engaging with Foursquare creatively in this way (Frith 2015c), users were thought to be able to "discover more outwardly legitimate places to immerse themselves in, while continuing an identity that revolves around a desire for authenticity" (Saker 2017, 945). Well aware of the contradictions that are at play, Saker argues:

What is of importance here isn't so much the actuality of the authenticity qua the authentic spatial experience experienced, but rather that participants feel Foursquare provides them with spatial information rooted in authenticity. That this feeling of authenticity *feels* authentic is what is important. (945)

It has also been suggested that the performance of an "authentic self" (and creating in readers a "perception" of authenticity) is just as important to Yelp users, applying equally to those leaving venue tips as to those reading them (Kuehn 2016). In a striking turn of phrase, Kathleen Kuehn views these practices as forms of "image entrepreneurship," a term she defines as an "economized construction of the self that deploys a favorable presentation of identity along a market-based logic" (Kuehn 2016, 2). While now more commonly associated with "micro-celebrity" (Senft 2008) and social media "influencers" (Abidin 2017), this is an idea that remains equally applicable to the ways that locative media users, as have been examined in this chapter, are enrolled in a "constant articulation" (Kuehn 2016) of their own sense of self-identity to their audience(s). As Lucy (student, aged 16–24) puts it, locative social media use is "definitely about crafting the kind of person that you want to present to people, like, you know, 'I've been to lots of cafes and bars,' like, you know, [I've] got lots of friends and [I'm] very social, and that sort of thing." Elsewhere in the same interview, the "image entrepreneurship" that is at play here is expressed even more explicitly: "So, if I like something that is interesting to say about where I was, or funny, or relevant, like, then I would send it. Something relevant to my personal brand" (Lucy, student, aged 16–24).

Third, the focus on sites of commerce in critiques of Foursquare neglects other, potentially richer lines of enquiry. While the various tactical uses of Foursquare check-ins that work to make space visible and legible to other users has been noted above and elsewhere (Humphreys and Liao 2013), how spaces remain *invisible* or *illegible* through services like Foursquare is less often remarked upon. An important exception is Jordan Frith's (2017) examination of "forms of spatial segregation" that result from the systematic exclusion and invisibility of certain venue types—such as the eateries popular among the Hispanic communities of Denton, Texas—within popular spatial search applications, like Yelp. In a related vein, there is also concern that the reliance on Foursquare to "orchestrate chance" while walking in the city risks entanglement "within an increasingly refined urban filter bubble that feeds back to the user what he or she feeds to the algorithms of and the networks of Foursquare" (van der Akker 2015; see also A. Crawford 2008).

It is for such reasons that Humphreys and Liao (2013, n.p.) call for detailed consideration of the "means through which those in power, e.g., Foursquare venues and Foursquare itself, exert power over the social construction of place through mobile social media." Humphreys (2012, 504) addresses this point elsewhere by noting the multitude of ways that "mobile social networks catalogue socio-spatial information about the users of these systems"—processes, as examined in detail in Part I of this book, that are "essential to the business of mobile social networking through the transformation of information into databases" (505).

CONCLUSION

This chapter has drawn on my own fieldwork and a now extensive literature on Foursquare end-use to examine how urban spaces and places are explored, catalogued, and communicated, and how these communicative practices are entwined with individual identity negotiation, performance, and display. What this work highlights is the need to take seriously, and continue to think through, the complicated identity work that is undertaken by users through locative social media and search and recommendation platforms *and* as a result of their own self-enrollment in commercially focused contexts of use.

What also clearly needs to be acknowledged is that these practices and platforms are by no means static and unchanging. Indeed, the mobile locative data landscape has changed significantly since the launch of Foursquare in 2009. Much of the novelty of venue check-ins associated with Foursquare, circa 2009, has dissipated. Foursquare itself, as discussed at length in Chapter 3, has transmogrified into a search and recommendation service, with social check-ins persisting in the company's less popular spin-off app, Swarm. At an individual level, contemporary "expressions of location and physical presence are now more creative and complex than proclaiming: 'I am here now'" (Halegoua et al. 2016, 10). In addition and crucially, beyond the level of the individual, there are now vast quantities of geocoded metadata accrued via an array of location-sensitive social media and search and recommendation services, "data that are constructed of both visual and textual information based on people's history of visitation" (Schwartz 2015, 92). This information is also subject to increasingly sophisticated algorithmic sorting, as well as interpretation that employs machine learning and AI techniques (Wilken 2018b). Yet, this information is not always presented back to users of these services (Humphreys 2011, 591–592). While the database and self-narrative functions of locative social media platforms are co-implicated, "complementary ordering structures," there inevitably remains a clear power differential privileging platform owners over end-users (van Dijck 2013c, 207). As these systems of data extraction, storage, and sorting multiply and expand, often in ways that are beyond end-user control, the potential regulatory and policy implications of, and end-user consumer concerns with, these systems magnify. It is to these considerations that discussion now turns in Part III of this book, where I take up complicated issues around how location data are being extracted at volume, by whom, and to what ends, and the many implications this has for individual consumers, as well as policy debates regarding privacy and data markets.

Geodata Capture and Privacy

Introduction

Part III, the final part of this book, takes up and explores a complicated set of issues relating to the following: how location data are being extracted, by whom, and to what ends; the many implications this has for individual consumers; and policy debates concerning data storage and privacy. Part III also marks a key shift in critical emphasis within the book. Whereas the latter two chapters of Part II, in the main, focused on smartphone applications where location is (or once had to be) actively registered by end-users, the arguments of Part III are centered around mass and automated location data-collection processes that often occur without the knowledge or beyond the control of end-users.

Chapter 7, "Location Data Extraction and Retention," examines two separate case studies that pertain to the acquisition and retention of various forms of geocoded data extracted from locative media devices and associated infrastructure, and where the political economic interests of governments and corporations are in tension. The first of these examines the controversies that flowed from revelations that Google had been gathering Wi-Fi data as part of its Street View operations. The second focuses on the US National Security Agency (NSA) and its far-reaching surveillance program, as revealed through the Edward Snowden papers. The chapter concludes by reflecting on the impacts of corporate and corporate-state data extraction and retention, and the legacies of the two specific cases under examination.

The final chapter of the book, Chapter 8, "Mobile Social Networking and Locational Privacy," explores one of the most pressing yet thorniest of policy issues pertaining to social media and mobile communications use for companies, end-users, and regulators alike: privacy. The chapter critically engages with the established and rapidly expanding body of work on privacy, examines how privacy implications and impacts are negotiated by consumers, and attempts to develop conceptual models of privacy that account for the reality of fast-moving traffic in ("decontextualized") "spatial big data" (Leszczynski and Crampton 2016).

The arguments of both of these chapters have been developed by drawing from a combination of relevant academic scholarship and other material, including financial statements, trade press reportage, and gray literature (Pappas and Williams 2011) such as unpublished corporate reports and industry analyses. These last two—trade papers and gray literature—form especially valuable resources for developing a deeper understanding of corporate practice as it relates to public policy (Corrigan 2018; Lawrence et al. 2015). In addition to these research sources, Chapter 8 also balances discussion of these sources with findings drawn from the same study of mobile location-based social media end-users detailed in the Introduction to Part II of the book.

CHAPTER 7
Location Data Extraction and Retention

Online interactions leave traces.
> —Humphreys (2012, 497)

Governments are not the only ones following users' locations; in fact they may be collecting far less than many corporations.
> —Landau (2016, 57)

Geolocation is of great interest to state actors and corporate entities because spatial data are inherently meaningful.
> —Leszczynski (2017, 240)

INTRODUCTION

In 2011, researchers from technology company O'Reilly Media, presenting at the company's Where 2.0 conference, revealed a piece of software they had developed called iPhoneTracker (Allan 2011). The iPhoneTracker software extracted location data from Apple's iOS4 operating system and overlaid it on a map, presenting clear traces of where an iPhone user was and had been. The release of this software caused major international controversy, and was a public relations nightmare for Apple (K. Crawford 2012, 223). The controversy wasn't due to the creation of the O'Reilly software so much as for what it revealed about the nature of Apple's actions:

> Here was a claim that Apple had installed a log file that was storing this sensitive data [the latitude and longitude of the phone's location] on every iPhone and backing it up with every synchronization [via iTunes to the user's computer

where it was stored unencrypted], without permission of users, and in such a way that it could be readily accessed. (K. Crawford 2012, 224)

While the log file was known to have existed in iOS3 (Levinson 2011), it was only with the upgrade to iOS4 that the data from these files began to be "recorded indefinitely" (K. Crawford 2012, 224). Apple was quick to claim that it was only tracking devices via cell tower base stations (and thus not with a high degree of location accuracy), and that the indefinite collection of data was a result of a bug (224). It did, however, admit to the collection of "anonymous traffic data [from users' movements] to build a crowd-sourced traffic database, aiming to give iPhone users improved traffic services" (224). What perturbed Apple users at the time of this controversy "was the permanent and unending nature of the data collection" (224). This was of concern given that "large volumes of data, collected over years, can reveal an incredibly detailed depiction of an individual's life, associations and preferences" (224; see also Leszczynski 2015).

The following year, researchers from Carnegie Mellon University and Rutgers University released results of a study they conducted testing users' expectations of what end-user information was captured by popular Android smartphone apps appearing in Android Market. What they found was a high degree of end-user surprise that apparently benign apps, like Brightest Flashlight, and mobile games like Angry Birds, and Toss It (where players toss a ball of crumpled paper into a waste bin), were collecting device ID and location data and sharing this information with third parties (mainly targeted mobile advertising and ad optimization firms) (Hong 2012; Lin et al. 2012; cf. Snow, Hayes, and Dwyer 2016). Angry Birds has since come to be considered as particularly problematic insofar as, in the words of one tech press article, it "leaks data like a sieve" (Leyden 2014). This assessment was based on a detailed analysis in 2014 by security vendor FireEye of end-user data traffic between the Angry Birds app and ad optimization firm Burstly. The FireEye researchers' analysis revealed a multistep procedure, whereby Angry Birds collects user's personal information that is associated with a customer ID, and this information is retained in the smartphone's storage. Then, "the Burstly ad library embedded in Angry Birds fetches the customer id, uploads the corresponding personal information [which includes location information] to the Burstly cloud, and transmits it to other advertising clouds" (Su, Zhai, and Wei 2014, n.p.).

In opening this chapter, I have presented just two among numerous controversies concerning the bulk extraction and storage of end-user location data that show this has been a long-standing issue (see Mims 2018).

The motivations behind this bulk collection of data are, over time, becoming increasingly clear.

The traces of our digital passage noted in the epigraphs to this chapter, and which have been documented in detail in the previous chapter and elsewhere in this book, are vital to the extraordinary growth, popularity, and commercial success of social media and search services like Facebook and Google. Users voluntarily sign up to mobile social networks. All the while, these "mobile social networks *catalogue* socio-spatial information about the users of these systems" (Humphreys 2012, 504; emphasis added). But it is not just social networks that catalogue end-user information. Rather, "just about every device is now connected and generating vast quantities of digital traces about interactions, transactions, and movements whether users are aware or not" (Ruppert, Isin, and Bigo 2017, 1). Thus, regardless of the source, cataloguing is now "essential to the business of mobile social networking through the transformation of information into databases" (Humphreys 2012, 505)—large-scale "storable and sortable" (Andrejevic 2010, 311) databases flush with rich, end-user-generated, geocoded data. These data are of great interest to corporations (especially ad targeting firms); these data are also of great interest to various government agencies, who have been "collecting, storing, retrieving, analysing and presenting data that records [*sic*] what people do and say on the Internet" (Ruppert, Isin, and Bigo 2017, 1).

We are often willing participants in this process. As opaque as terms and conditions documents tend to be, by signing up to and using social media and search services, we are trading access to and use of these corporate services for corporate permission to access and use data about us and our patterns of engagement and our behaviors.

However, not all forms of data collection operate according to this broadly reciprocal arrangement; there have been numerous cases where end-user data have been accessed, accrued, and stored outside of end-user knowledge and consent, and which push up against the boundaries of accepted regulatory and legal frameworks.

This chapter, one of two in the third and final part of this book, explores two such cases in detail. The first of these cases is on Google's Street View location-capture service and the controversies that followed its rollout. In particular, I will focus on the damaging revelations that Google's Street View cars were collecting and storing personal Wi-Fi data, and the legal and regulatory responses that followed across a number of different national markets and jurisdictional contexts. This first case was selected as it was regarded internationally at the time as a serious breach of privacy, and an example of significant corporate overreach in terms of the types of personal data that were being gathered and the means by which these data were obtained.

The second case focuses on Edward Snowden's revelations concerning the US National Security Agency (NSA) and its far-reaching surveillance program. This second case was selected for its enduring impact. Not only did it generate significant international furor at the time, and brought to light vivid details concerning massive (corporate-assisted) government surveillance, it has also, as will be argued at the end of the chapter, shifted the terms of policy debate concerning geolocation data capture, data security, corporate-government surveillance, and corporate social responsibility. In looking at the NSA revelations, my focus is firmly on the centrality of geolocation to these data capture operations. Geolocation data are of particular importance to security agencies because they are *spatio-temporal* (they record information about space and place and movement through spaces and places) and *spatio-relational* (they record information about our relationships with others in/and places) (Leszczynski 2015, 969). The chapter concludes by reflecting on a number of issues that are raised by both of these cases.

GOOGLE STREET VIEW AND WI-FI DATA COLLECTION

In mid-2007, Google—already the clear market leader in search and in consumer-facing location services (Google Maps, Google Earth)—sought to further strengthen its position with the launch of Google Street View. This is a service that captures panoramic 3D street-level images of streetscapes in major cities and towns in the United States and internationally. Street View pictures are acquired from a fleet of cars equipped with roof-mounted cameras providing 360-degree image capture, and the data are stored in onboard computers. These images, once uploaded into Google's servers, are then interfaced with and augment Google's Search and Maps facilities.

Street View Pictures Concerns

Much of the initial consternation generated by Street View concerned the perceived privacy impacts associated with a large multinational corporation recording public and private spaces, and people going about their day-to-day activities, without prior consent. Responses to these concerns differed across jurisdictions, and included everything from calls for revision of opt-in/opt-out arrangements, and pixilation of certain images, to lowering the height of car-mounted Street View cameras, and, in some cases, temporary ceasing of operations, as well as fines and the threat of fines. Widespread concern also resulted in the generation of a significant body of legal scholarship exploring the privacy impacts and possible regulatory and legal responses to Street View (see, for example: Geissler 2012; Lavoie 2009; McGowen 2010; Meese and Wilken 2014; Rakower 2011; Segall 2010; Strachan 2011; van der Sloot and Zuiderveen 2012; Wiggers 2011). For other critics, privacy issues were by no means the only concern. As Scott McQuire (2016, 79) argues, "the most significant shift pioneered by Street View is not its generation of photographs of identifiable individuals, but how it enables the wholesale conversion of urban space into data." By doing this, McQuire suggests, Google "effectively *appropriated* something—the public appearances of urban space—that previously belonged to no one, and it has converted this common resource into private value" (McQuire 2016, 84, emphasis in original; Alvarez León 2016). In making this conversion, Google's aim is to further enrich their "database of user mobility patterns, preferences and purchasing behaviour" (Hoelzl and Marie 2014, 267).

Street View Wi-Fi Data Controversy

In mid-2010, further controversy erupted following revelations that, in addition to photographing streets, Google's Street View cars had also been capturing location information and Wi-Fi data.

Concerns over the forms of data that Google's Street View cars were collecting were first raised by Johannes Caspar, the data protection supervisor in Hamburg, Germany. Caspar had been leading the German government's liaison efforts in working to address concerns raised by the Street View service. Writing in the *New York Times*, Kevin O'Brien (2010) reports that, in May 2010, Caspar inspected a Street View car at the invitation of Google. Caspar apparently noticed that the hard drive had been removed from the car and requested to inspect it, but was told that this wasn't possible as the data were encrypted (K. O'Brien 2010). From Caspar's persistent questioning about what data had been captured via Street View cars, Google agreed to an inspection of the hard drive.

The results of this hard drive inspection were published on Google's Europe Blog in a post written by Peter Fleischer (2010), Google's European-based Global Privacy Counsel. In his post, Fleischer explained that Google Street View cars collected three kinds of information: street view photos, 3D building imagery, and local Wi-Fi network data. All of these sources of information were considered vital to Google's larger location services efforts. The first, street-level photographs, were taken in order to build Street View, and "to improve the quality of [. . .] maps, for example by using shop, street and traffic signs to refine [. . .] local business listings and travel directions" (Fleischer 2010, n.p.). The second, 3D building imagery, or "3D geometry data," were collected using "low power lasers" in order to "help us improve our maps" (Fleischer 2010). The third, Wi-Fi network data, were collected in order to "improve location-based services like search and maps" (Fleischer 2010). What Google meant by "Wi-Fi network data," Fleischer's post explained, was information that identified the network. This information included SSID (service set identifier) data, which are 32-character Wi-Fi network name identifiers, and MAC (media access control) addresses, which are unique numbers given to devices like Wi-Fi routers.

Fleischer's mention of Wi-Fi network data collection was significant in that it was the first acknowledgment by Google that its Street View cars were conducting "wardriving." "Wardriving" is the practice of "searching for Wi-Fi wireless networks" from a moving vehicle:

> Wardrivers use a Wi-Fi equipped device together with a GPS device to record the location of wireless networks. [. . .] The maps of known network IDs can be used as a geolocation system—an alternative to GPS—by triangulating the current position from signal strengths of known network IDs. (Wikipedia 2017b, n.p.)

Google's revelations of "wardriving" did not please European government representatives. Germany's Federal Commissioner for Data Protection, Peter Schaar, claimed that he had not been made aware of this when permission was granted for Google to take pictures for Street View; a similar statement was made by Peter Day, a spokesperson for the UK Information Commissioner's Office, who was quoted as saying that "at no point did Google make us aware that it would be scanning Wi-Fi too" (cited in Marks 2010, n.p.).

Fleischer (2010, n.p.) admitted that, with the benefit of hindsight, "greater transparency would have been better." However, in divulging that the Google Street View program collected Wi-Fi network data, Fleischer was at pains to stress four things: (1) that this network data did not identify households or individuals; (2) that "Google does not collect or store payload data"; (3) that all three forms of data Google collected via its cars were also being collected by other firms operating at that time, including NavTeq, TeleAtlas (pictures and 3D building imagery), and Skyhook Wireless (Wi-Fi network data); and (4) that the collection of Wi-Fi network data was not illegal, and that is why their use of "wardriving" wasn't disclosed. Each of these statements would prove significant in the context of later revelations; and, with the exception of the third, each was later challenged.

On May 14, 2010, Google released a second blog post, this one penned by Alan Eustace, Google's then senior vice president of Engineering and Research. In this post, Eustace declared that the statement made in Fleischer's post of a few weeks earlier—that "Google does not collect or store payload data"—was in fact incorrect. "It's now clear," Eustace (2010b, n.p.) wrote, "that we have been mistakenly collecting samples of payload data from open (i.e. non-password-protected) WiFi networks, even though we never used that data in any Google products." This "mistake" was attributed to the work of a single engineer who, in 2006, was "working on an experimental WiFi project [and] wrote a piece of code that sampled all categories of publicly broadcast WiFi data" (Eustace 2010b).

In response, Eustace (2010b) announced that Google had appointed a data analytics firm, Stroz Friedberg, to conduct a third-party assessment of the Street View software source code (known as "gslite"), "with particular focus on the elements of wireless network traffic that the code captured, analyzed, parsed, and/or wrote to disk" (Stroz Friedberg 2010, 1). "Parsing" in this context refers to the process of "reading in a data stream of some sort and building an in-memory model or representation of the semantic content of that data, in order to facilitate performing some kind of transformation on the data" (Garrison 2012, n.p.).

From their analysis of the gslite source code, Stroz Friedberg confirmed that Google had in fact been collecting MAC addresses and SSID data, which it then associated with longitude-latitude coordinates obtained from a GPS unit located in the Google Street View vehicle. The report also explained that gslite "works in conjunction with an open source network and packet sniffing program called Kismet" (Stroz Friedberg 2010, 2). In accessing wireless network traffic, gslite worked in conjunction with Kismet to capture and store to a hard drive "the header information for both encrypted and unencrypted wireless networks" (2). Header data are the explanatory information that appear at the beginning of a block of stored or transmitted data (Wikipedia 2017b). Payload data are what follow header data—that is, they are the actual intended contents of a message (Wikipedia 2017b). Stroz Friedberg (2010, 2) refer to payload data as the "body of wireless data packets," and this is where "e-mails or file transfers, or evidence of user activity, such as Internet browsing, may be found."

With respect to payload (or "body") data capture, the report found that gslite did not retain payload data transmitted over encrypted wireless networks. Unencrypted networks were a different matter, however, with gslite writing to a hard drive "the bodies of wireless data packets" from these networks; it was claimed that the Google Street View computers did not "attempt to analyze or parse that data" (Stroz Friedberg 2010, 2).

How did Google manage to capture Wi-Fi-related data? When a Google Street View car traveled along a residential street, it not only collected street view photos and 3D building imagery, it also listened for wireless routers. During this process, the onboard computer recorded network identifying information (SSID data and MAC addresses) that were matched to GPS coordinates. In addition, the Street View software (gslite, working with Kismet) saved non-password-protected Wi-Fi data being transmitted across the networks the car encounters. As the car passed by, the onboard computer shifted between 11 different Wi-Fi network frequency bands, or "channels," every 2.2 seconds. This gave the car's computer roughly one-fifth of a second to connect to a Wi-Fi network. In this time, depending on whether the car was stationary or moving and at what speed, the data collected from each network—header data in the case of encrypted networks, and header and payload data in the case of unencrypted networks— amounted to around 250 kilobytes, "equal to roughly 25 e-mails or more" (King and Gröndahl 2012, n.p.). The brevity of access to each Wi-Fi network led Google's Alan Eustace (2010b) to claim that Street View cars would "typically have collected only fragments of payload data."

The Street View payload data capture revelations sparked investigations in at least 17 different countries, including South Korea—which has strict data privacy laws (Y.-C. Kim 2012; Tweney 2013) and restrictions over the provision and use of maps data (Geens 2012; Pfanner 2013)—where the cybercrime unit of the Korean National Police Agency (KNPA) raided Google's Seoul office, seizing computers and hard drives (Halliday 2010a; Song 2011). (For a timeline and full details of these investigations, see https://epic.org/privacy/streetview/.)

In a number of jurisdictions, including Canada, France, and the Netherlands, samples of payload data were analyzed. In Canada, an examination of a sample of payload data by the Office of the Privacy Commissioner of Canada (OPC) "revealed, among other information, the full names, telephone numbers, and addresses of many Canadians" (OPC 2011). The investigation also revealed that "complete email messages, along with email headers, IP addresses, machine hostnames, and the contents of cookies, instant messages and chat sessions" (OPC 2011, n.p.). Especially concerning to the OPC (2011) was the discovery of "particularly sensitive information, including computer login credentials (i.e., usernames and passwords), the details of legal infractions, and certain medical listings." The OPC (2011) found that "the information collected was sufficiently capable of being linked to individuals through data matching or aggregation." In France, the

Commission Nationale de l'Informatique et des Libertés (CNIL) identified data relating to internet navigation (including passwords) and further data relating to electronic mail, including information that identified individuals and matched them to specific residences with a high degree of accuracy (CNIL 2011; Ellison 2012, 12–13). And, in the Netherlands, the Dutch Data Protection Authority (DDPA) concluded from their analysis of payload data that the "recorded data are not meaningless fragments" and that it was "factually possible to capture 1 to over 2,500 packets per individual user in 0.2 seconds" (cited in Ellison 2012, 13). Furthermore, the DDPA discovered that "Google had captured a broad assortment of Internet traffic, including e-mails, chat traffic, URLs, passwords, and video and audio files, some of which was highly sensitive" (Ellison 2012, 13).

In the United States, the Federal Communications Commission (FCC) also investigated Google Street View Wi-Fi data collection. While the FCC did not gain access to sample payload data, their investigation reviewed Google Street View design documentation, and internal company correspondence. These documents yielded important additional information: that the engineer responsible for Street View Wi-Fi capture, referred to in FCC-Google dealings as "Engineer Doe" (but subsequently named in later legal proceedings), was instructed to add wardriving capabilities during the design stages of Street View (Ellison 2012, 10); that Wi-Fi data extraction was not an "accident" and that code was "deliberately written to capture payload data" (10); and that the decision to scrape Wi-Fi header and payload data was not the action of a single "rogue" engineer but was undertaken with the knowledge of other members of his team, including Street View project leaders, who were aware of the processes involved and the forms of data being obtained and stored (14–18; see also Orlowski 2012). In debunking Google's "it was an accident" claim, the FCC investigator, P. Michele Ellison observed:

In a discussion of "Privacy Considerations," the [Google Street View] design document states, "A typical concern might be that *we are logging user traffic* along with sufficient data to precisely triangulate their position at any given time, *along with information about what they were doing*." That statement plainly refers to the collection of payload data because MAC addresses, SSIDs, signal-strength measurements, and other information used to map the location of wireless access points would reveal nothing about what end users "were doing." (Ellison 2012, 11)

Finally, while the FCC acknowledged the Stroz Friedberg finding that Google did not store data from encrypted networks, the FCC investigator did make a point of noting that a submission made by Google to international data protection authorities included "a vague statement about Google's ability to determine whether a Wi-Fi network is, or is not, unencrypted" (Ellison 2012, 5, fn. 33). The FCC observation opens the door to the possibility that the delineation between encrypted and unencrypted may not have been so neat as previously argued.

Following these investigations by external regulators, Google's Alan Eustace (2010a, n.p.) amended his earlier statement regarding the collection of payload fragments to read: "It's clear from those inspections that while most of the data is fragmentary, in some instances entire emails and URLs were captured, as well as passwords." In response, Google apologized and promised to remove Wi-Fi sniffing devices from its Street View cars and delete stored payload data. Despite their apologies and these assurances, throughout the entire Wi-Fi data collection saga, Google continued to deny any legal wrongdoing, insisting that the Wi-Fi data they captured were publicly accessible.

Crime and Punishment

Across various jurisdictions, there was general consensus that Google had broken the law in covertly extracting and retaining Wi-Fi data. In Australia, Google was deemed to have contravened the Privacy Act, with Federal Communications Minister Stephen Conroy describing Street View as the "single greatest breach in the history of privacy" (Halliday 2010b, n.p.). The Canadian Privacy Commissioner described Street View's Wi-Fi data collection as a "serious violation of Canadians' privacy rights" (Halliday 2010b, n.p.), and it was also concluded that Google had broken relevant privacy laws in Germany. In the United States, however, the FCC ruled (on the basis of insufficient evidence) that Google did not breach the US Wiretap Act (Ellison 2012, 23), while the Federal Trade Commission (FTC) also conducted its own investigation without taking action against Google (Ellison 2012, 8).

Across the board, however, international regulatory and legal responses to Google's Wi-Fi data collection indiscretions could be viewed as little more than saber-rattling, with the punishments meted out to Google largely only symbolic (for detailed discussion of international legal outcomes, see Burdon and McKillop 2013; Chow 2013). In Australia, for instance, Privacy Commissioner Karen Curtis instructed Google to publish a public apology, and to undertake a "Privacy Impact Assessment (PIA) on any new Street View data collection activities in Australia that include personal information," but admitted that, under the current Australian Privacy Act, she was "unable to impose a sanction on an organisation" when she had initiated the investigation. Rather, her role, she stated, was to "work with the organisation to ensure ongoing compliance and best privacy practice" (OAIC 2010, n.p.). In Canada, Google was instructed to comply with certain security recommendations and delete all confidential data captured there (Halliday 2010b). In France, Google was fined €100,000, while Ireland and Britain did not impose fines after Google agreed to delete data collected in those countries (C. Miller 2013). In Germany, Google was fined €145,000, close to the maximum fine under EU law of €150,000—which, it has been pointed out, "amounts to 0.002% of Google's $10.7 billion net profits in 2012" (RT 2013, n.p.). In the United States, the FCC ruled that Google did not breach the US Wiretap Act, but fined Google US$25,000 for hindering FCC investigations (Ellison 2012, 23). Google subsequently agreed to pay "a fine of $7 million and police its own employees on privacy issues after a lawsuit brought about by 38 states" (RT 2013, n.p.).

Google's Wi-Fi Data Collection Motives?

Rarely discussed in tech and mainstream press coverage throughout the Google Wi-Fi saga were Google's motives for wanting to collect Wi-Fi network information and Wi-Fi payload data. Wi-Fi network information improves Google's location-based services, including those offered through its Google Geo Location API. Google's Peter Fleischer (2010, n.p.) described a possible scenario where a Google Maps for Mobile user enabled "My Location" in order to "identify their approximate location based on cell towers and WiFi access points which are visible to their device." Thus, Google's efforts to gather MAC addresses and SSID data further enrich its database of existing Wi-Fi access points (Sterling 2010). And "by recording of the location and ID of wireless routers, Google can triangulate a smartphone user's location faster and with less power than using satellites" (King and Gröndahl 2012, n.p.). In this way, the Google Street View "wardriving" collected Wi-Fi network data that contribute in vital ways to Google's master or base map, known as "Ground Truth" (McQuire 2016, 80). Meanwhile, Wi-Fi payload data add contextual richness for Google by revealing user "habits, preferences, routes and routines"—information that is "critical to the data-hungry economy" (McQuire 2016, 79) and vital to Google's geodemographic profiling efforts (Barraneche 2012a). In broader business terms, these combined streams of geolocation data are core to Google's commercial success. As Alex Madrigal (2012, n.p.) notes,

"Google's geographic data may become its most valuable asset. Not solely because of this data alone, but because location data makes everything else Google does and knows more valuable."

A more specific motivation for Google's Wi-Fi data collection program was to gain clear competitive advantage over the key rival firm, Skyhook Wireless. Skyhook Wireless was an early pioneer of mobile location services. Prior to the launch of Street View, Skyhook had one of the most comprehensive databases of Wi-Fi access points, populated, by its own open admission, through wardriving (Jones 2015). Skyhook had also developed proprietary location determination software (Skyhook XPS) for mobile handsets, mobile geofencing technology, and various geodemographic analysis services (SkyhookWireless 2017). In light of these developments, Google's decision to engage in wardriving to capture location and payload data from unencrypted Wi-Fi channels was seen as part of their "competitive imperative to create a more accurate mobile-location-service than Skyhook" (Cleland 2010, n.p.).

At the same time as the Street View Wi-Fi data collection controversy was unfolding, Google also discovered that mobile handset makers Motorola and Samsung had signed agreements with Skyhook Wireless (on top of those already signed with Google) for the provision of mobile location services. Google executives were quick to realize that OEMs (original equipment manufacturers) switching to Skyhook would be "awful for Google, because it will cut off our ability to continue collecting data to maintain and improve our location database" (cited in Patel 2011). Understanding that "control over core data is critical" (McQuire 2016, 79), especially given Google's investment in the Android operating system (Lunden 2011) and the rapid rate of uptake of Android-powered devices, Google acted swiftly and aggressively in response, instructing Motorola and Samsung to withdraw from their deals with Skyhook, which both OEMs subsequently did. Google's somewhat spurious argument for this withdrawal was that Skyhook XPS software failed Android OS compatibility tests and that Skyhook XPS data would "contaminate" Google's location database, and confuse users, by registering "Skyhook WiFi/cell tower locations as GPS locations" (Patel 2011, n.p.).

What followed were legal proceedings brought by Skyhook Wireless against Google over two issues: (1) anti-competitive behavior (with Skyhook arguing that Google had bullied Motorola and Samsung to pull out of signed agreements); and (2) Google infringement of four location-related Skyhook patents (Hachman 2010). In 2014, the first case was dismissed as Google was deemed to be "within its rights to demand that Samsung and Motorola break their deals with Skyhook, because of the two companies' pre-existing contracts with Google" (Bray 2014, n.p.). In the second case, in which Skyhook was "alleging that Google had illegally used its cellphone location technology, which relies on Wi-Fi stations instead of GPS satellites" (Newsham 2015a, n.p.), the two firms reached settlement just prior to trial, with Skyhook receiving US$90 million in damages (Newsham 2015b).

Samson Esayas (2017, 143) makes the pertinent observation that "the growing importance of personal data for commercial purposes is fueling the desire to collect and amass as much personal data, both through legitimate (eg acquisitions) and illegitimate (eg deliberate deception) mechanisms." Google's Street View Wi-Fi data collection program falls firmly within the latter category. Driving Google's voracious appetite for personal spatial data was the belief that, when it comes to large data sets, "the sum is more valuable than its parts, and when [one] recombine[s] the sums of multiple datasets together, that sum too is worth more than its individual ingredients" (Mayer-Schönberger and Cukier 2013, 108). While it would be erroneous to suggest that Google's determination to push Skyhook Wireless from Android location service provision was the *sole* motivation behind its hoovering up of Wi-Fi data, it certainly played a key part.

The outcome of this whole messy episode can be viewed in two ways. On the one hand, the Street View data-collection program, as well as the Skyhook lawsuits, appear to have been quite a high-stakes gamble for Google, given the fallout from the former and the US$90 million fine they were handed in the latter. On the other hand, perhaps the stakes have not been so high after all? Google's financial punishment for its global Wi-Fi data-collection program has been minimal to modest, at best, and, more significantly, it has since emerged with secure longer-term control of vital location resources on Android devices, and a position as the clear global corporate leader of consumer-oriented location services. What is striking is that, amidst all the legal and regulatory mess and noise of the Wi-Fi data scandal, this longer-term strategic play has received remarkably little coverage, and appears to have succeeded largely unchallenged.

In the second half of this chapter, I turn my attention from the Google Street View Wi-Fi data-collection controversy to a rather different (but not altogether unrelated) case—that concerning Edward Snowden's revelations about the US National Security Agency's (NSA) data-surveillance programs.

THE NSA AND LOCATION DATA EXTRACTION AND RETENTION

In early June 2013, the first of former intelligence analyst and NSA contractor Edward Snowden's many explosive revelations about the covert surveillance operations of the US government's NSA came to light. For much of the remainder of 2013 and into 2014, the global mainstream press cycle was dominated by revelations, published by the *Guardian*, the *Washington Post*, and other outlets, drawn from the trove of documents that Snowden had amassed about the NSA's many secretive intelligence-gathering surveillance operations.

What was so striking about the Snowden files at the time of their release was not that these documents drew wider public attention to the NSA. Indeed, the existence of, and the covert surveillance work undertaken by, the NSA had been known for a long time, and has been well documented in a series of trade press books by writer James Bamford (1982, 2001, 2008). Rather, what was considered so revealing about the Snowden files was the light they shed on the scale and voraciousness of the NSA's global communications data-gathering programs; the involvement of major US tech and telecommunications companies; the fact that the NSA's data-gathering operations extended well beyond security priorities to include diplomatic and trade-related interests; the extensive partnerships with other international intelligence agencies that enabled data-sharing and, through this, the circumvention of rules preventing domestic surveillance; and, particularly concerning for the US press, the fact that NSA programs included domestic surveillance of US citizens.

These all were and remain significant issues. However, given the concerns of this book, the focus here is on tracing the crucial importance of geolocation data to the NSA's operations. The documents released by Snowden revealed that geolocation information was either the explicit focus of what was being gathered, or was an intrinsic component of innumerable other, larger data-gathering operations. In tracing the vital importance of geodata to the NSA's surveillance programs, my intention here is not to rehearse the circumstances around the release of the Snowden files (on this, see Greenwald 2014), nor is it to provide a comprehensive account of the chronology of events (for detailed, cross-referenced timelines, see Gidda 2013; Szoldra 2014). Rather, my aim here is to revisit key NSA programs that are principally concerned with geodata capture (either in whole or in significant part), exploring what forms of geodata were extracted, how access was obtained, how these data were stored and then analyzed, and to what ends. What the Snowden documents reveal, Agnieszka Leszczynski (2015, 968) argues, is that the "mining and capture of locational data are not incidental to the capture of other personal information"; rather, these data are of central concern to the NSA because of their sensitivity (not least in identifying and situating individuals) (966–967).

Interspersed throughout this analysis are illustrations by artist Trevor Paglen, who has sought to document pictorially various aspects of the NSA's operations. These images serve as part visual guide, part visual counterpoint to the at times dizzying array of textual information on these programs, and part anchor-point, giving some semblance of a tangible visual presence to otherwise seemingly intangible data-surveillance operations.

Before turning to a consideration of the NSA's interest in geodata, it is valuable to begin by considering the main actors involved. David Lyon (2014, 3) suggests that the NSA's capacity to intercept and capture communications data at scale would not have been feasible without the involvement of three sets of actors: ordinary users of digital media and communication technologies, private corporations, and government agencies. Ordinary users unknowingly disclose information when using social networking platforms, when using phones, and, as will be discussed later, through the NSA's gleaning of communications metadata (Lyon 2014, 3). All of this end-user information is "sucked up as data, quantified and classified, making possible real-time tracking and monitoring" (4). Obtaining this rich and prized end-user data would not be possible, though, without the cooperation—willing or otherwise—of telephone, internet, and web companies. Following the release of the Snowden files, it also emerged that tech companies were required to provide access to the NSA (Greenwald 2014, 108ff.), but they weren't compelled to make access easier (112), even though some elected to do so (113ff.). David Lyon (2014, 1) has referred to this as "the ambiguous complicity of internet companies," and there are recent suggestions that there is (especially in Google's case) a long, convoluted, and often secretive history to this complicity (Ahmed 2015a, 2015b; Levine 2015). Indeed, Esayas (2017, 157) goes as far as to say that "surveillance by private entities is the engine that powers government surveillance." In addition to telco and tech companies, the NSA also routinely engaged other corporations as contractors, such as Booz Allen Hamilton (Snowden's employer prior to him fleeing to Hong Kong and then Russia), in order to "share the burden of their work" (Lyon 2014, 2) in gathering and mining vast quantities of user data. Finally, in addition to drawing data from certain corporations and seeking analytics assistance from others, there was the work undertaken by the NSA (and other government agencies) to develop systems and programs in order to extract, store, and analyze this vast trove of globally sourced data. It is these last programs that I wish to explore in detail in the following.

One of the striking aspects of the Snowden files is the staggering array of covert NSA programs with enigmatic code names, which the artist

Trevor Paglen sought to collect and document in his "Code Names" exhibition (see Figure 7.1; and see Angwin and Larson 2014 for an interactive chart of known NSA-related coded secret programs).

Many of these code names—"Evil Olive," "HappyFoot," "NoseySmurf," "EgotisticalGoat"—would be comical were it not for the seriousness of what they entailed. One of Snowden's major accomplishments was to "decode" these programs by piecing together information that he had been systematically amassing over many years, and which had been gleaned from a variety of classified sources (Greenwald 2014, 90ff.).

The Snowden files revealed not only that geolocation data were a crucial concern across virtually all of the NSA's secretive surveillance programs Snowden gathered information on, but that these data were and are being collected on an unprecedented scale. Barton Gellman and Ashkan Soltani (2013b) report on how global cell phone location data—in excess of 5 billions records a day—were fed into a vast database known as "Fascia." The contents of this database, a NSA white paper revealed, were analyzed using "a powerful suite of algorithms, or data sorting tools" (Soltani and Gellman 2013). One such set of tools, known as "Co-traveler Analytics," was developed by the NSA's "Five Eyes" surveillance partner, the Australian Signals Directorate. "Co-traveler Analytics" examined "mobile Call Detail

Figure 7.1. Trevor Paglen, *Code Names*, 2016.
Source: Photograph by Raimund Zakowski. Reproduced with permission.

Records containing location information" in order to predict potential points of connection (co-presence) between individuals who happened to be traveling in close physical proximity to, or who follow similar mobility patterns, to a moving surveillance target (Soltani and Gellman 2013). The principle of the "Co-traveler Analytics" process is as follows: the number of nearby mobile phones traveling in close proximity to a mobile phone belonging to a person of interest is likely to decrease quite rapidly as the target individual moves from one mobile phone cell area to another; if other mobile phone users are still nearby to the targeted phone user after passing through four to six cell phone areas, they, too, are likely to become persons of interest.

The data feeding these storage and analytics tools were drawn from a wide array of sources, with the NSA targeting mobile and internet communications and their supporting infrastructures (mobile networks, cell towers, internet company servers, data centers and web traffic, Wi-Fi signals, terrestrial and submarine fiber-optic cable networks, and satellite communications). Key NSA programs were designed for the explicit purpose of extracting geolocation data.

NSA PROGRAMS SPECIFICALLY TARGETING GEOLOCATION DATA

Each time a mobile device communicates with a cellular network, it conveys vital pieces of information about the location of that device. Each time a mobile phone connects to cellular networks, it immediately registers its location in one or more telecommunications-related databases, which are known as HLRs (Home Location Registers) and VLRs (Visitor Location Registers). The VLR assigns a TMSI (Temporary Mobile Subscriber Identifier) to every mobile device in the area the moment it is switched on. In addition, each mobile device has a number of other specific identifiers: an IMEI (International Mobile Equipment Identifier), a unique number that is given to every single mobile phone; an IMSI (International Mobile Subscriber Identifier), a unique identifier that defines each subscriber, including the country and the mobile network to which the subscriber belongs; and a MSISDN (Mobile Station International Subscriber Identity Number), which is the unique phone number, including the country code, of a subscriber. The Snowden files revealed that the NSA was particularly eager to capture the preceding and related forms of information, including DNR (Dialed Number Recognition) and DNI (Digital Network Intelligence) data (Greenwald 2014, 132; Soltani and Gellman 2013), and developed a range of ways to capture and extract this information. For instance, it established "Juggernaut" to exploit a perceived vulnerability in the SS7 network, which is a "global data protocol that links phone networks together," in order that it could "process raw feeds" transmitted between mobile carriers (Soltani and Gellman 2013, n.p.). It also developed "HappyFoot," a program whose primary focus was geodata extraction. "HappyFoot" involved the interception and bulk collection of "traffic generated by mobile apps that send a smartphone's location to advertising networks" (Soltani and Gellman 2013), and of location-based data transmitted through search and recommendation apps that draw on GPS signals, cell towers, or Wi-Fi signals to determine precise location (Soltani, Peterson, and Gellman 2013).

Many of the NSA's efforts to obtain smartphone location data were revealed by Snowden to have been conducted with the cooperation of the UK equivalent of the NSA, the Government Communications Headquarters (GCHQ) (Ball 2014a). The GCHQ reportedly developed an entire family of smartphone-related targeting tools, with the title of each one modeled on character names from the kids' cartoon *The Smurfs*. TrackerSmurf was developed for tracking precise geolocation (there was also NoseySmurf for remotely turning on a phone's microphone, and DreamySmurf for remotely activating a turned-off device, among others) (Ball 2014a).

In addition to the TrackerSmurf tool, the two agencies sought to extract data in bulk from a variety of smartphone apps. One way they did this was to target so-called leaky apps like Angry Birds (Hill 2010), targeting them for their location and user profile data (Serrano 2014). Another even more sophisticated effort involved collecting large volumes of location data by "intercepting Google Maps queries made on smartphones" (Ball 2014a), and by extracting the geographic metadata that are embedded in photos "when someone sends a post to the mobile versions of Facebook, Flickr, LinkedIn, Twitter and other services" (Glanz, Larson, and Lehren 2014, n.p.). The Google Maps process was said to be so successful that a GCHQ document from 2008 noted that it "effectively means that anyone using Google Maps on a smartphone is working in support of a GCHQ system" (quoted in Ball 2014a), and a NSA report from the previous year boasted that, based on intercepts of global searches for directions, "you'll soon be able to clone Google's [places] database" (Glanz, Larson, and Lehren 2014).

Both agencies also worked on mapping existing telecommunications infrastructures by building a database for "geolocating every mobile phone mast in the world" (Ball 2014a, n.p.). Potentially, this would enable the NSA and the GCHQ to glean location information about a mobile handset based solely on the ID information of the cell towers that device interacted with (Ball 2014a). These joint efforts also seemed to share some similarities with the disclosure, made by NSA director Keith B. Alexander during a Senate testimony in October 2013, that the NSA ran a pilot project run in 2010–2011 to test the feasibility of collecting US cell phone tower and location data (Gellman and Soltani 2013b; Savage 2013).

Finally, one of the more unexpected discoveries to emerge subsequent to the release of the Snowden files concerned the NSA's penchant for filing patents with the US Patent and Trademark Office (Corona 2015; Harris 2014). The NSA, it has been revealed, holds almost 300 patents (Corona 2015); these are generally kept from public view, until somebody files an identical patent (Corona 2015). Two of these patents were for geolocation identification systems (Huffman, Spring, and Reifer 2002; D. Smith 2011). The second of these, with the title "Device for and Method of Geolocation," is particularly interesting in light of the earlier discussion of the NSA's "Co-traveler" program. This patent seeks to protect a NSA-developed means of determining the location of a transmitter based on its interactions with receivers (see Figure 7.2).

The NSA's patent application states: "the difference in radial velocities and delay time between the signals received at the receivers are used to geolocate the transmitter" (D. Smith 2011, n.p.). This NSA patent describes a process that is also remarkably similar to now established methods of determining the location of a mobile device via cell tower triangulation and GPS trilateration. Although it is unclear, precisely, how the NSA's patented process differs from established means of doing this, the fact that the NSA lodged such patents was clearly an attempt to try to protect the technical methods they had established in order to be able to geolocate mobile signals.

Other NSA Programs Capturing Geolocation Data

In addition to the aforementioned programs that target geolocation information explicitly, there were a raft of related programs revealed in the Snowden files that also sought to capture geolocation information as part of larger data-collection efforts. Here I mention a number of the more significant of these, which the NSA used to capture information via mobile communications, internet traffic, undersea cables, satellite communications, and through various forms of metadata.

With respect to the targeting of mobile communication, the NSA had at least two specific programs. The first, "Dishfire," was concerned with the "untargeted collection and storage of [global] SMS messages" (Ball 2014b, n.p.); these messages were used to gather credit card details, contact lists,

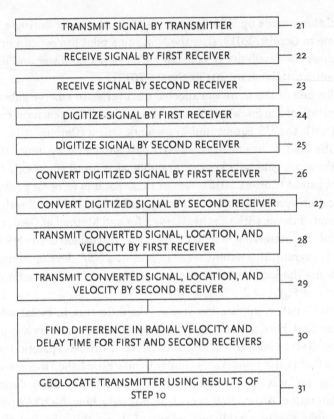

TRANSMIT SIGNAL BY TRANSMITTER	— 21
RECEIVE SIGNAL BY FIRST RECEIVER	— 22
RECEIVE SIGNAL BY SECOND RECEIVER	— 23
DIGITIZE SIGNAL BY FIRST RECEIVER	— 24
DIGITIZE SIGNAL BY SECOND RECEIVER	— 25
CONVERT DIGITIZED SIGNAL BY FIRST RECEIVER	— 26
CONVERT DIGITIZED SIGNAL BY SECOND RECEIVER	— 27
TRANSMIT CONVERTED SIGNAL, LOCATION, AND VELOCITY BY FIRST RECEIVER	— 28
TRANSMIT CONVERTED SIGNAL, LOCATION, AND VELOCITY BY SECOND RECEIVER	— 29
FIND DIFFERENCE IN RADIAL VELOCITY AND DELAY TIME FOR FIRST AND SECOND RECEIVERS	— 30
GEOLOCATE TRANSMITTER USING RESULTS OF STEP 10	— 31

Figure 7.2. Diagram illustrating NSA patented "Device for and method of geolocation," 2009. *Source:* D. Smith (2011).

and "location information about where a user has been" (Spiegel 2013, n.p.). (The GCHQ also created "Thieving Magpie" in 2012, a program designed for GSM [Global System for Mobile Communications], to intercept and monitor onboard mobile phone services on aircraft [Greenwald 2014, 164–166].) While "Dishfire" was largely concerned with capturing metadata, the second program, "Mystic," more ambitiously recorded actual phone calls, "storing billions of them in a 30-day rolling buffer that clears the oldest calls as new ones arrive" (Gellman and Soltani 2014). This program, which had the capacity, it has been claimed, to record "100%" of a foreign country's phone calls, captured information on the names, dates, and *locations* of these calls (Gellman and Soltani 2014). Recorded snippets of messages that were deemed to be of enduring interest could be retained, and these were retrievable using the NSA's "Retro" search facility. This tool was considered most valuable when an NSA analyst uncovered a new phone number or name of interest; armed with this information, the analyst could use "Retro" to "pull an instant history of the subject's movements, associates and plans" (Gellman and Soltani 2014, n.p.).

Then there were the NSA's internet traffic-related data-gathering programs. Of these, perhaps the best known is "PRISM," a program that provided the NSA with "front door" access (approved by secret legal process) to the "central servers of nine leading US Internet companies" (Microsoft, Yahoo, Google, Facebook, PalTalk, AOL, Skype, YouTube, and Apple) in order to extract "audio and video chats, photographs, e-mails, documents, and connection logs," as well as rich geocoded data (Gellman and Poitras 2013, n.p.; see also MacAskill and Rushe 2013). The NSA also developed the "Muscular" program to facilitate "backdoor" (covert and unapproved) access to the data centers of Yahoo and Google (Gellman and Soltani 2013a). (A related UK GCHQ pilot program, "Squeaky Dolphin," was run to monitor and access Google and YouTube data in real time [Esposito et al. 2014].)

In addition to "PRISM," the NSA created numerous programs to target wider internet traffic. Several of these ("Fairview," "Stormbrew," "Blarney," "Oakstar") were concerned with so-called upstream surveillance—accessing "communications on fiber cables and infrastructure as data flows past" (NSA slide, quoted in Greenwald 2014, 108). These were all part of the NSA's ambitions to "own the Internet," according to former NSA senior executive and whistleblower Thomas Drake (cited in Kloc 2013).

One way the NSA achieved access to fiber optic cables was via secretive deals with telecommunications firms, such as its long-term arrangement with AT&T (Angwin et al. 2015), and through related deals struck by its international intelligence partners, like that between BT, Vodafone Cable, and Verizon Business (among others) that gave the GCHQ access to undersea cables (Ball, Harding, and Garside 2013). Another way this was achieved was via secret intelligence gathering arrangements between the NSA, its other "Five Eyes" partners, and 33 of what it calls "third-party" countries, including Denmark, Germany, and 15 other European Union member states (R. Gallagher 2014). These deals, which were part of the NSA's "RAMPART-A" program, allowed the NSA to tap into fiber optic cables at various "congestion points around the world" and provided the NSA with direct access to "phone calls, faxes, e-mails, internet chats, data from virtual private networks, and calls made using Voice over IP software like Skype" (R. Gallagher 2014, n.p.), as well as "'vast volumes' of location data from around the world" (Gellman and Soltani 2013b, n.p.). Olga Khazan (2013) explains that intelligence agencies such as the NSA "likely gain access to the landing stations" for undersea cables—such as that at Marseille, France, photographed by artist Trevor Paglen in Figure 7.3, as part of his project to document these landing stations—and "usually with

Figure 7.3. Left: Trevor Paglen, *NSA-Tapped Fiber Optic Cable Landing Site, Marseille, France*, 2015. Right: Trevor Paglen, *Approaching Marseille*, 2015.
Source: Trevor Paglen. Reproduced with permission.

the permission of the host countries or operating companies." The NSA then used small devices "to capture the light being sent across the cable" (Khazan 2013, n.p.). With these devices, known as "intercept probes," the NSA bounced the light that was traveling through the cables through a prism; the NSA then "makes a copy of it, and turns it into binary data without disrupting the flow of the original Internet traffic" (Khazan 2013).

While "RAMPART-A" focused on fiber optic cables, "Echelon," a further program run by the NSA and its "Five Eyes" partners, and arguably one of the most secretive, specifically targeted satellite communications. Largely hidden listening stations, such as that at Sugar Grove, West Virginia (since closed) (see Figure 7.4)—sites that artist Trevor Paglen refers to collectively as the "dark geography" of surveillance (Gustafsson 2013, 159; Paglen 2008)—permitted the NSA and partners to intercept and record incoming international communications messages (Bamford 2005). While the existence of the "Echelon" program has long been known, the Snowden files revealed further details about it, including the fact that it was built in the Cold War era, not to spy on Russia and China and their allies, but rather "to target Intelsat satellites, which in the early years were used primarily by Western countries" (Campbell 2015).

The Snowden files also shed light on an airborne system designed by the NSA for capturing communications information at scale. This was achieved through an aircraft-mounted platform called "Shenanigans" for vacuuming up enormous amounts of data from wireless routers, computers, smart phones, or other electronic devices that happened to be within range (Scahill and Greenwald 2014). One six-month mission, code-named "Victorydance," a joint NSA-CIA operation run out of the CIA's Oman drone base, consisted of 43 flights over the Arabian Peninsula that successfully "mapped the Wi-Fi

Figure 7.4. Photographic image of the NSA's Echelon program listening facility at Sugar Grove, West Virginia. This facility has since been closed. Trevor Paglen, *They Watch the Moon*, C-print, 36 x 48 inches, 2010.
Reproduced with permission.

fingerprint of nearly every major town in Yemen" (quoted in Scahill and Greenwald 2014). (Wi-Fi data, of course, as has been discussed at length in the first part of this chapter and elsewhere in this book, provide crucial information for accurately positioning mobile devices.)

Finally, the NSA had also developed various large-scale programs whose sole purpose seemed to be the omnivorous capture of metadata. One of these, "Marina," was a facility that "aggregates NSA metadata from an array of sources" (Ball 2013, 2014c). Another, "X-Keyscore," Glenn Greenwald (2014, 153) described as both a storage facility (the NSA's "'widest-reaching' system for collecting electronic data") and a search facility (where searches could be conducted based on "email address, telephone number, or identifying attributes such as an IP address" and location). And then there was "BoundlessInformant," a tool for cataloguing global surveillance data that could be used, according to an internal NSA document, to "map and view metadata volume" and other details by country (Greenwald and MacAskill 2013).

What holds all of these geolocation data extraction efforts together— "the software, the algorithms, the codes that allow users' data to be systematically extracted or disclosed, [and] analysed" by the NSA—is the promise that this "intelligence" can be translated into "actionable data" (Lyon 2014, 3). Thus, it is perhaps unsurprising that geolocation data of the sort described in the preceding was equally crucial to the US military's drones program (Scahill and Greenwald 2014), and "central to target development" (Leszczynski 2015, 969). Unmanned aerial vehicles (UAVs), or drones, "piloted" by US military's Joint Operations Command (JSOC) drone operators, use coordinates provided by the NSA's Geo Cell tracking program (Scahill and Greenwald 2014). A key means by which they could home in on drone targets was by "analyzing the activity of a SIM card, rather than the actual content of the calls" made using that SIM card (Scahill and Greenwald 2014). SIM card and cell phone information was accrued by intercepting communications, from cell towers, via ISPs, and by hacking multinational mobile phone chip maker Gemalto in order to obtain encryption keys (Scahill and Begley 2015). In addition, the NSA also equipped its Predator drones, which were already fitted with thermal infrared sensors for locating targets (Parks 2014), with "devices known as 'virtual base-tower transceivers'—creating, in effect, a fake cell phone tower that [could] force a targeted person's device to lock onto the NSA receiver without their knowledge" (Scahill and Greenwald 2014). These devices were fitted as part of the NSA's "Gilgamesh" program, which the NSA described as Predator drone-based "active geolocation" (cited in Scahill and Greenwald 2014). These "virtual base-tower transceivers" permitted the US military to track a cell phone "to within 30 feet [9 meters] of its actual location, feeding the real-time data to teams of drone operators who conduct missile strikes or facilitate night raids" (Scahill and Greenwald 2014, n.p.). Geo-tracking of this kind was performed in collaboration with the US's "Five Eyes" alliance partners, including staff stationed at Australia's Pine Gap signals facility (Dorling 2013).

The Gilgamesh program attracted strong criticism from critics opposed to the covert military use of armed drones for assassinations. The SIM-based method of geolocation is open to "gaming" by those targeted who seek to create confusion by "purposively and randomly" distributing and swapping SIM cards and by carrying multiple SIM cards (Scahill and Greenwald 2014); there has also been the deployment of additional tactical measures to evade/confuse the drone's other sensors (Parks 2014). As a result, drone strikes based on this method are regarded as notoriously imprecise. Not only are they said to result in "death by unreliable metadata" (Scahill and Greenwald 2014), these extrajudicial killings—as data artist Josh Begley explored in his Metadata+ project discussed in Chapter 4—routinely involve the deaths of people beyond those targeted. Indeed, the Bureau of Investigative Journalism (Bureau 2017) in the United Kingdom reported that, between August 2004 and August 2017, the US military had conducted a minimum of 428 confirmed drone strikes in Afghanistan, Pakistan, Yemen, and Somalia, resulting, conservatively, in 2,511–4,020 total deaths; of these deaths, 424–969 were civilians, and 172–207 were children.

The Changed Tense of Surveillance Post-9/11

In describing all of the NSA's aforementioned geolocation-focused programs, it is important to remember that intelligence gathering by governments—particularly the interception of communications signals (SIGINTEL)—is by no means new. Abby Hinsman (2014) details how the CIA had been trialing aerial photographic spy planes during the early 1950s. And Glenn Greenwald has described how, in a US context, the FBI's domestic counterintelligence program (COINTELPRO), which was uncovered by the Senate Church Committee in the 1970s, also revealed that, between 1947 and 1975, the NSA had access to millions of private telegrams sent from, to, or passing through the US "under a secret arrangement with three United States telegraph companies" (Greenwald 2014, 185). There is also a view that the NSA's embrace of wiretapping and covert data capture was connected to its long-standing, yet ultimately failed, interest in controlling encryption technologies (see Diffie and Landau 2007; Landau 2016).

What the Snowden files make very clear, however, is that things have shifted significantly over the past few decades, with intelligence gathering and surveillance occurring at an unprecedented scale and level of sophistication, such that it now includes the targeting of whole populations. As Lyon (2014, 4) writes, "the political-economic and socio-technical responses to 9/11 helped to change the 'tense' of surveillance in [. . .] significant ways." Whereas, in the past, targeting involved identifying a suspect and then seeking data about him or her, this process largely occurs in reverse:

> Now bulk data are obtained and data are aggregated from different sources *before* determining the full range of their actual and potential uses and mobilizing algorithms and analytics not only to understand a past sequence of events but also to predict and intervene *before* behaviors, events, and processes are set in train. (Lyon 2014, 4)

This shift in "tense" has also been enabled by a "wholesale technological transformation of a process once limited by its materiality" (Weld 2014,

38). Whereas once the question was "how many holes could fit on a punch card, how many pieces of mail could be intercepted" (Weld 2014, 38), now advances in computing have permitted the construction of programs like "Silverzephyr" (that collects individuals' phone and internet metadata at scale) and "XKeyscore" (the NSA's vast database that is searchable by name, email, IP address, language, and geolocation).

It would be naive, however, to think that the NSA's systems are not affected by their own significant infrastructural constraints. James Bamford (2012), for example, writes of the NSA's push to develop a vast, secretive data center in Utah with the capacity to store communications data in perpetuity, and how the NSA's desire for increased analytic capacity is driving a new kind of "space race": that focused on the development of ever-faster supercomputing power.

Driving this is the desire for predictive analytics and "the quest for pattern-discovery" (Lyon 2014, 1). This desire was used as a key justification for the NSA wanting bulk access to data, and was reflected in former head of the NSA Keith B. Alexander's personal motto: "collect it all" (Greenwald 2014, 95). "Bulk methods," Gellman and Soltani (2014, n.p.) write, "capture massive data flows. [. . .] By design, they vacuum up all the data they touch."

IMPACTS OF STATE-CORPORATE DATA EXTRACTION AND RETENTION

For geographer Jeremy Crampton (2015, 524), "the most surprising element of the geopolitical assemblage revealed in the Snowden documents has been the public-private cooperation of government and Internet companies." The cumulative impacts of the NSA Snowden revelations have included a shift in public sentiment toward corporations and governments and a change in the tenor of debate around what Dencik, Hintz, and Cable (2016, 1) refer to as "state-corporate surveillance," or "dataveillance."

In the immediate aftermath of the Snowden revelations, loss of trust in the state and its agencies and policies was viewed by political commentators as a significant issue. Indeed, at his year-end press conference in 2013, US president Barack Obama defended NSA surveillance as vital to the United States' national security interests, but admitted that there was significant work to be done, on the national stage, in rebuilding public trust in US government policies, and, on the international stage, in rebuilding the trust of its long-time allies, including France, Germany, and Spain (Clayton 2013).

Internet firms were also concerned about loss of trust among consumers and the reputational impacts of their involvement in the NSA's PRISM program. The nine internet companies, which included Facebook, Apple, Google, Microsoft, and Yahoo, responded by denying that they provided the NSA with unfettered access to their servers; implored the US government to rein in the NSA; and fought for greater transparency about the PRISM program and the right to explain publicly what sorts of information they were required to hand over to government agencies (Facebook Reveals 2013). A number of these tech firms have also since moved to improve and strengthen encryption (Landau 2016, 59).

Within academic discourse, there has been increased attention paid to the clear power asymmetries that are at play, both in relation to broader practices of "big data" extraction and retention, and in specific relation to the NSA and its programs. The Snowden documents provided evidence that demonstrated "the intricate ways in which everyday communication [has been] integrated into an extensive regime of surveillance that relies considerably on the 'datafication' of many aspects of social life" (Dencik, Hintz, and Cable 2016, 1). If one approaches datafication and data "as an object whose production [and collection and storage] interests those who exercise power" (Ruppert, Isin, and Bigo 2017, 3), then a key issue becomes that of data access and control. Data monitoring at scale invariably favors the few (who hold power) over the many (who do not) (Dencik, Hintz, and Cable 2016, 10). With the rise of "big data," and in the wake of the release of the Snowden files, there has emerged a wealth of work that seeks to respond to and redress these power asymmetries. Much of this work falls under the rubric of "data politics." As Evelyn Ruppert, Engin Isin, and Didier Bigo (2017, 1) define it, "data politics asks questions about the ways in which data has become such an object of power and explores how to critically intervene in its deployment as an object of knowledge." Under this banner, to cite just two examples, there have been efforts at developing the concept of "data justice," which proposes a "reframing [of] data debates to consider how digital infrastructures and data-driven processes have implications for broader society beyond individual privacy" (Dencik, Hintz, and Cable 2016, 9), and exploring the possibilities of "representative data management" to address some of the limitations of "data privacy self-management" (Obar 2015, 13), which has become largely untenable in the face of "spatial big data" (Leszczynski 2015).

Then, in legal, policy, and regulatory discourses, there has been a great deal of attention paid to "state-corporate surveillance" and corporate-sponsored "dataveillance" and their privacy impacts; such is the importance of privacy to our understandings of locative media that it will be examined at length in the next chapter. In relation to state-corporate surveillance, Glenn Greenwald (2014, 94) concludes that "[t]aken in its entirety, the

Snowden archive led to an ultimately simple conclusion: the US government had built a system that has as its goal the complete elimination of electronic privacy worldwide." (This is a claim that has also been leveled against social media firms and echoes past statements by Facebook's Mark Zuckerberg and Alphabet's Eric Schmidt on the declining social importance of privacy.) What does it mean for our conceptions of privacy, particularly at a national and international scale, when one nation can intercept and store the entire telephony traffic of another nation?

The question of privacy came to the fore in public debate around the significance of bulk metadata collection by the NSA. Metadata has been described by the NSA as "information about content (but not the content itself)" (Greenwald 2014, 132). In the United States, the collection of metadata was permitted after 9/11 under the Patriot Act. And, in response to the Snowden revelations about metadata collection, the US Department of Justice refers to a 1979 US Supreme Court ruling to consistently defend its position by "categorizing metadata as business records in which there is no expectation of privacy" (Clayton 2013). In addition, the US government has sought to argue that metadata collection is not intrusive, "or at least not to the same degree as intercepting content" (Greenwald 2014, 133).

Scholars like Ed Felton have challenged this position vociferously. In an affidavit filed by the American Civil Liberties Union challenging the legality of the NSA's metadata collection program, Felton paints a compelling picture of how individual pieces of metadata, when combined, can build a strong narrative about an individual that effectively serves as a "proxy for content" (see Greenwald 2014, 134). As Greenwald illustrates,

> Metadata about an email message, for instance, records who emailed whom, when the email was sent, and the location of the person sending it. When it comes to telephone calls, the information includes the phone numbers of the caller and the receiver, how long they spoke for, and their locations and the types of devices they used to communicate. (Greenwald 2014, 132)

The clear point here is that "metadata is surveillance" (Bruce Schneier quoted in Lyon 2014, 3), and, as such, carries clear implications for the identification and the privacy of individuals.

In light of the Snowden revelations, there have been moves toward achieving improved forms of data protection (the EU's Global Data Protection Regulation being a case in point; see McDermott 2017). There have also been calls for increased understandings of data and agency (H. Kennedy, Poell, and van Dijck 2015) and greater digital literacies (McCosker 2017) to arrest what Jonathan Obar (2015, 1) refers to as "a perpetual information illiteracy" experienced by consumers in the face of big spatial data.

The NSA's location programs (by their own admission) also generate so much information that it is "outpacing [their] ability to ingest, process and store" data (Gellman and Soltani 2013a, n.p.), creating what have been termed "unwieldy heaps" that NSA computer programs struggle to sort and analyze (Glanz, Larson, and Lehren 2014). Given this, there is merit in revisiting academic digital memory debates addressing the extraction, storage, and use of personal location information. Work in this area advocates for (among other things) an "ethics of forgetting" (Dodge and Kitchin 2007; Blanchette and Johnson 2002; see also Green 2009), and the inclusion of sunset clauses so that stored personal data is deleted after a certain time period (Mayer-Schönberger 2009).

In addition to the previously mentioned initiatives and opportunities, many significant further questions remain to be answered. Key among these, for Glenn Greenwald (2014), is the issue of a lack of oversight—in Greenwald's words, "the NSA is the definitive rogue agency: empowered to do whatever it wants with very little control, transparency, or accountability" (131). There is also the issue of "scope creep" or overreach ("great quantities of the program manifestly had nothing to do with national security," 94), with many surveillance targets "plainly economic in nature" (135), leading Jeremy Crampton (2015, 519) to conclude that "there is no clear distinction between the strategic and economic."

Meanwhile, for Jeremy Crampton, a geographer, the Snowden revelations are, in geolocational terms, merely the tip of the iceberg. Crampton (2015) makes the point that, while the Snowden files "provide perhaps the single largest insight into Big Data surveillance, they are mostly limited to one agency (NSA)" (522), and there are in fact 16 agencies within the US intelligence community (520), at least two of which specifically relate to the gathering of geolocation intelligence. One of these, the National Geospatial-Intelligence Program (NGP)—"the remote sensing imagery, GIS, and geographic intelligence (GEOINTEL) effort of the US government" (Crampton 2015, 522)—is mentioned in the Snowden documents, where it is revealed that it consumes 9% (or US$4.7 billion) of the US intelligence budget, employing 8,500 people (Gellman and Miller 2013). In addition to the NGP, Crampton also makes mention of the National Geospatial-Intelligence Agency (NGA), and the National Reconnaissance Office (NRO), which take care of US surveillance satellites (Crampton 2015, 521). For Crampton, much more is required if we are to understand the operations of these agencies and their contractors (which include GIS companies such as Esri).

There is, however, a twofold difficulty in pursuing any of the concerns described in the preceding analysis. The first, as David Lyon (2014, 5) points out, is that much discussion is "occurring in a [policy] context [particularly in the US, but also elsewhere] that celebrates rather than carefully assesses Big Data" extraction and retention, and which views such data capture as a necessary means to preventing future terrorist attacks (McDermott 2017, 4). As Greenwald (2014, 74) reminds us, the 2008 US FISA (Foreign Intelligence Surveillance Act) Amendments Act sought to institution-alize rather than halt the warrantless wiretapping that occurred under the Terrorist Surveillance Program introduced by US president George W. Bush after 9/11. The second is that these discussions are also occurring in a con-text where much of the data that are being extracted and retained are "'volunteered' data" generated by end-users of mobile devices, wearables, and social media platforms, even though "the users of such devices may not think of themselves as volunteering data to others" (McDermott 2017, 4). What is more, while privacy (the right to be let alone) presupposes some form of intrusion, the Snowden documents reveal that much "data-veillance" now occurs completely unbeknownst to end-users (McDermott 2017, 3).

CONCLUSION

In this chapter, I have explored two key and controversial cases—Google's Street View Wi-Fi data collection, and the geolocation information cap-ture efforts of the NSA—where end-user geolocation data have been accessed, accrued, and stored outside of end-user knowledge and consent, and which push up against the boundaries of accepted regulatory and legal frameworks. The first of these cases, it could be said, has fueled consumer distrust of and government concern (especially within the EU) over the corporate reach of internet firms, and the lengths they sometimes go to, and the means they are prepared to employ, in order to populate their databases with end-user-generated geolocation data. The second of these cases has not only changed the tense of surveillance (according to David Lyon), it has also changed the tenor of debate around state-corporate trust, data protection, and privacy. Reflecting on the impact of the Snowden revelations five years later, Ewen MacAskill and Alex Hern (2018) note that public awareness has been raised, new legislation has been passed in both the United States and the United Kingdom, encryption has become commonplace, and the disclosures concerning the extent of collaboration between the intelligence agencies and internet companies hastened the arrival of the European Union's data protection legislation, which took effect in May 2018.

All of the issues raised in this chapter point to the ongoing importance of privacy. Privacy has long been and continues to be a crucial issue in wider popular engagement with and critical understanding of location-based services. Taking up the question of privacy in the next chapter, I examine in greater detail and depth how the concept of privacy has been and continues to be understood, how individual locative media end-users negotiate their own understandings of location privacy through everyday practice, as well as exploring some of the conceptual and practical privacy challenges posed by the emergence of "spatial big data" (Leszczynski 2015).

CHAPTER 8
Mobile Social Networking and Locational Privacy

Your location history is a signature of your life, and if you see that you can figure out a lot more.

—Ted Morgan, Skyhook Wireless CEO, cited in Lunden (2011)

Our personal spatial data flow freely and without friction across and between interoperable and synergistic geo-enabled devices, services, applications, and analytics engines.

—Leszczynski (2017, 236)

INTRODUCTION

In this chapter, I take up and examine one of the thorniest issues facing location-sensitive mobile social media end users, corporations, and regulators: privacy. The analysis of privacy in this chapter proceeds in three steps.

First, I explore some of the definitional challenges presented by the concept of privacy. I also detail how geolocation data capture and sharing bring further privacy complications to our use of smartphones and social media applications.

Second, I examine how these privacy implications and impacts are negotiated by consumers, by drawing on findings from an interview-based study of locative media end-users in Melbourne, Australia, and New York City (see Introduction to Part II for details). What is revealed from these data are nuanced individual methods for seeking to control the generation and dissemination of personal locational data—methods that, at the same time, are often riven with contradictions and compromises. These intricate, if often contradictory, practices raise a number of important questions that this part of the chapter aims to explore: Why do consumers express anxiety about locational data sharing, yet forego some degree of privacy control when they sign up to social networking and location-based services for personal benefit? How might we make sense of apparent discrepancies between personal declarations about the importance of privacy, actual privacy settings, and practices?

And, in the third and final part of the chapter, I examine conceptual accounts of privacy that attempt to come to grips with the reality of the fast-moving traffic in "decontextualized" location data that move within and between corporations and other bodies as a result of cross-platform partnerships, data-sharing arrangements, and so on. Here I argue that Helen Nissenbaum's (2010) influential model of "privacy as contextual integrity" remains arguably the most sophisticated attempt to date to develop a theory of privacy that accounts for the realities and challenges posed by "spatial big data." Yet, as I go on to point out, Nissenbaum's model is not without criticism. The argument of this final section is that, increasingly, established privacy models are viewed as no longer adequate or sufficient in addressing the complications and complexities associated with "spatial big data." Indeed, we appear to be at a particularly important critical juncture as we attempt to grapple with the challenges posed by, and appropriate means of responding to, these spatial big data flows.

UNDERSTANDING PRIVACY

One of the challenges facing any analysis of privacy is coming to grips with the manifold meanings that are associated with the term. Privacy has been described as an "evanescent" (Gormley 1992, 1336), "elusive" (Trottier 2018, 472), "contested" (Hartmann 2011, 191) concept, and a "messy and complex subject" (Nissenbaum 2010, 37). It has been labeled as "a concept in disarray" (Solove 2006, 477), and an "elastic" (Margulis 2011, 14) and "overextended concept" (Hartzog and Selinger 2013, n.p.) for the ways that is quite often used, particularly in the tech and popular press, to capture a variety of concerns, ranging from being observed or disturbed by others, being identified by others, breaches of confidentiality, fears about various forms of surveillance, to being evoked "to discuss issues associated with corporate access to personal information" (Hartzog and Selinger 2013), and to express concerns about data security. Some explanation for the increased definitional wooliness around privacy will be offered toward the end of this chapter. My immediate aim in the discussion to follow, though, is to provide a summary of some of the pioneering and prevailing approaches to the conceptualization of privacy.

In their foundational essay, Samuel D. Warren and Louis D. Brandeis (1890) define privacy, and the protection of the person (207), as the "right to be let alone" (193)—that is, privacy as "freedom from surveillance and scrutiny" (Trottier 2018, 470). Following this seminal account, there have been innumerable attempts to further refine definitional understandings of privacy. For example, Daniel Solove (2006) develops a taxonomy of privacy in "an attempt to understand various privacy harms and problems that have achieved a significant degree of social recognition" (483). Here, however, I focus on Ken Gormley's (1992) survey of one hundred years of privacy literature, which is useful for understanding the variety of ways that privacy has been understood in legal and social sciences scholarship. From his survey, Gormley argues that, throughout the twentieth century, privacy definitions have multiplied, yet have tended to cluster into four major categories. These are as follows:

(1) Privacy "viewed as an expression of one's *personality* or *personhood*," with these definitions focusing upon the right of the individual to define his or her essence as a human being" (Gormley 1992, 1337; original emphasis; see, for example, Pound 1915; Freund 1975);

(2) Privacy as marking "the boundaries of *autonomy*," and "the moral freedom of the individual to engage in his or her own thoughts, actions and decisions" (Gormley 1992, 1337; original emphasis; see, for example, Henkin 1974);

(3) Privacy as "citizens' ability to *regulate information* about themselves, and thus control their relationships with other human beings" (Gormley 1992, 1337–1338; original emphasis), an understanding of privacy that is most commonly associated with the work of Westin (1967), to be discussed further later, among others (e.g., C. Fried 1968); and

(4) Privacy formulations that take "a more noncommittal, mix-and-match [definitional] approach, breaking down privacy into two or three essential components, such as Ruth Gavison's 'secrecy, anonymity, solitude'" (Gormley 1992, 1338; see Gavison 1980), or regarding privacy, as Daniel Solove (2008, 98) does, as lacking a "unitary value" and consisting, instead, of "a plurality of protections against different types of problems."

It is the third of these categories that has arguably exerted the greatest influence on how we continue to approach privacy in the late twentieth and early twenty-first centuries, and particularly in relation to computer-mediated communication and social media use. Four influential approaches to thinking about privacy—by Alan Westin, Irwin Altman, Sandra Petronio, and Helen Nissenbaum—can all be situated within this third category. According to Alan Westin's influential definition,

> Privacy is the claim of individuals, groups, or institutions to determine for themselves when, how, and to what extent information about them is communicated to others. Viewed in terms of the relation of the individual to social participation, privacy is the voluntary and temporary withdrawal of a person from the general society through physical or psychological means, either in a state of solitude or small group intimacy or, when among large groups, in a condition of anonymity or reserve. (Westin 1967, 7)

Thus, privacy, according to Westin, is determined socially. Moreover, privacy is considered a "dynamic process (i.e., we regulate privacy so it is sufficient for serving momentary needs and role requirements) and as a non-monotonic function (i.e., people can have too little, sufficient, or too much privacy)" (Margulis 2003, 412; see Westin 1967, 32–42).

For Irwin Altman, like Westin, privacy is also viewed as dynamic and driven by what individuals are prepared to disclose about themselves; privacy, for Altman, is a "boundary regulation process involving [the] opening and closing of the self to others" (Altman 1990, 236) through "the selective control of access to the self" (Altman 1975, 24). Altman's conception of privacy is significant, according to Stephen Margulis (2003, 419), in that it gives emphasis to how privacy is "inherently a social process" that comes about through "the interplay of people, their social world, the physical environment, and the temporal nature of social phenomena," and that it has a cultural context ("all cultures have evolved mechanisms by which members can regulate privacy, but the particular patterns of mechanisms may differ across cultures" [Altman 1977, 70; Altman 1975]).

Then there is Sandra Petronio, whose work is said to have been stimulated by Altman's dialectical conception of privacy (Margulis 2011, 12). Petronio has developed a conception of privacy known as communication privacy management (CPM) theory. As Lee Humphreys and I have explained elsewhere (Humphreys and Wilken 2015, 296–297), CPM theory makes explicit the role of control in managing privacy, and one of the key principles of CPM theory is that people feel that they own their personal information and that, when they share it with others, the recipients become co-owners of this information. When we share personal information with others, Petronio (2002) argues, we become vulnerable because we lose some control of that information. The management of informational boundaries thus requires ongoing attention regarding the revealing and concealing of information by all co-owners. CPM theory suggests that individuals and collectives develop and engage in boundary management regarding the flows of private information (Petronio 2002, 2007). CPM theory argues that the co-ownership of information leads to a potential vulnerability, as one gives up solitary control over that information. And, in order to manage the vulnerability of co-ownership, Petronio (2002) suggests that actors engage in privacy rule foundations—that is, the processes by which the rules for revealing and concealing private information are developed. Control over information and boundary play—two processes that are central to Petronio's CPM theory—are also key dimensions of Christena Nippert-Eng's conception of privacy (Nippert-Eng 1996, 2005) and her understanding of how individuals tend to manage privacy in their everyday actions and activities as the interplay of "selective concealment and disclosure" (Nippert-Eng 2010, 2ff.).

Finally, there is Helen Nissenbaum's (2010) "contextual integrity framework." Two things sit at the heart of this framework. The first is the dual importance of *context* and *norms,* or what Nissenbaum (2010, 129) terms "context-relative informational norms." This is the idea that "expectations of privacy turn on norms" (Solove 2007, 167), on established understandings of what is appropriate in given situations or contexts. The second is "transmission principles." As Nissenbaum explains,

> A transmission principle is a constraint on the flow (distribution, dissemination, transmission) of information from party to party in a context. The transmission principle parameter in informational norms expresses terms and conditions under which such transfers ought (or ought not) to occur. (Nissenbaum 2010, 145)

Privacy, according to this framework, then, "does not mean the indiscriminate control of personal information, but a highly differentiated practice of sharing and withholding information depending on its meaning and sensitivity in different contexts" (Debatin 2011, 50). Consequently, as Debatin (2011, 50) goes on to explain, "violations of privacy are seen as violations of contextual integrity," or, as Nissenbaum (2010, 140) terms them, "breaches of context-relative informational norms."

The contextual integrity framework is significant for internet research, Nissenbaum argues, in that it is said to sidestep some of the difficulties that follow from conceiving of privacy around a public-private tension and the concomitant issue of sharing personal information online. As Nissenbaum (2010, 187) writes:

> If a right to privacy is a right to *context-appropriate flows* [. . .] there is no paradox in caring deeply about privacy and, at the same time, eagerly sharing information as long as the sharing and withholding conform with the principled conditions prescribed by governing contextual norms.

In this section of the chapter, I have documented one key attempt at categorizing different ways of understanding privacy (by Gormley), and have sketched a number of influential approaches to, or models of, privacy (by Westin, Altman, Petronio, and Nissenbaum). What is significant about the latter in the context of this chapter is that each of these models has had a significant impact in shaping debate about privacy within internet and mobile social media research (see, for example, Bazarova 2012; de Souza e Silva and Frith 2012; Humphreys and Wilken 2015; Margulis 2011; Marwick and boyd 2014; Salter 2016; Taddicken and Jers 2011; Waters and Ackerman 2011). Given the specific concerns of this book, it is also important to clarify the place and significance of location technologies to present understandings of privacy.

UNDERSTANDING LOCATIONAL PRIVACY

Within the location-based services industry, privacy has come to be regarded as a central yet especially sensitive issue. In an interview from 2011, Ted Morgan, at that time CEO of Skyhook Wireless, described privacy as the "third rail of location services" (cited in Lunden 2011). "Third rail" is a term drawn from (mainly US) political discourse that references something that is especially sensitive or controversial, and thus difficult to broach. (This sensitivity is also revealed by Apple's decision in 2012, ahead of the release of iOS6, to move location from general settings to the privacy section of settings.) Privacy has gained this reputation in relation to location services, as critical geography and communication and media studies scholars have explained, for at least three reasons.

The first reason has to do with the growing frequency with which personal identifying information—especially location data—has been compromised or mishandled. Such instances include the extraction of geolocation and other data from social media services without full consent; or, where potential system flaws that could lead to leaks, breaches, and privacy invasions have been exposed by researchers (e.g., the iOS4 data logs case discussed in opening the preceding chapter) and by others with privacy awareness raising pedagogical aims (see Leaver and Lloyd 2015). There has also been a proliferation of apps and practices that display varying degrees of "leakiness and creepiness" (Shklovski et al. 2014). These range from the "leakiness" of apps like Brightest Flashlight and the mobile game Angry Birds that, as discussed in the previous chapter, were collecting device ID and location data and sharing this information with third parties, such as targeted mobile advertising and ad optimization firms (Hong 2012; Lin et al. 2012; see also Mims 2018; Whittaker 2017), to the "creepiness" of apps like GirlsAroundMe that, circa 2010, drew data from Foursquare's and Facebook's APIs to display location information about nearby women who publicly shared their Foursquare and Facebook check-ins with users of the GirlsAroundMe app (Brownlee 2012; Frith 2015b, 129). In addition, there are also cases involving various forms of "digital deception" (Frith 2015b, 126), including the manipulation of geolocation information by so-called bad actors (such as those, for example, who use dating apps like Tinder in conjunction with "GPS spoofing" to fake or obscure one's precise location; see Zhao 2015). All of these examples have been highlighted by others in order to raise public awareness of, and increased skepticism toward, the manifold ways that social media services capture, manipulate, and exploit location data (see, for example, Khanna 2015).

The second reason privacy is regarded as the "third rail" of location services is because "many of our quotidian digital media practices are spatially oriented" and "depend on the availability of geocoded information as functional inputs" (Leszczynski 2017, 235) in order for them to operate successfully, as this book has explored at length (see also de Souza e Silva and Frith 2010, 2012). Thus, the growing "ubiquity and ordinariness of locationally enabled devices" (Leszczynski 2017, 235) means that, each and every day, we generate vast amounts of "spatial content as intended outputs or byproducts of our interactions with spatial media, at times unbeknown to us" (235).

And the third reason is due to the fact that the disclosure of location information is seen as "uniquely sensitive in terms of the kinds of personal details that it reveals" (Leszczynski 2015, 966; Hartmann 2011). Agnieszka Leszczynski (2017) provides the following explanations as to why the disclosure of location information is regarded as so sensitive:

(a) Personal spatio-temporal location data are seen to "constitute definitive proof or evidence of individuals' involvement in specific behaviours, activities and events in space, or [are] seen as proof of the potential for their involvement" (239).

(b) "The extensive, exhaustive and continuous nature of geosurveillance [. . .] means that there is not a feasible way of achieving or maintaining complete spatial anonymity within data flows" (239).

(c) As the first of the two epigraphs to this chapter makes clear, "spatial-relational data are inherently meaningful beyond being merely locational, revealing other intimate aspects of our personal lives" (239), particularly when "correlated with other kinds of personally identifying information (PII)" (236), including web cookies (Elmer 2004, 119) and other forms of metadata.

(d) Moreover, "unlike other forms of PII, spatial data carry with them information that can be used to translate threats to our personal safety and security into actual harms to our person" (Leszczynski 2015, 239).

All of the preceding points suggest that, with the rise of smartphones and mobile computing, location can be regarded as distinct from other forms of data within network contexts (Leszczynski 2015). As Adriana de Souza e Silva and Jordan Frith put it, once a "user's location becomes a crucial determinant of the type of data accessed," then, consequently, "privacy issues become more directly interconnected with location"—this they term "locational privacy" (de Souza e Silva and Frith 2012, 118; see also Blumberg and Eckersley 2009). With these specific concerns as background, in the following section, I turn to an examination of how locational privacy has been negotiated by consumer end-users of location-sensitive mobile social networking services.

PRACTICING PRIVACY

In light of the concerns expressed in the preceding section, it is no surprise to find then, as Kelli Burns (2013) has pointed out, that privacy has long been a key research consideration across much of the available literature on mobile and locative social media (see de Souza e Silva and Frith 2010, 2012; Farman 2012; Frith 2015b; Hartmann 2011; Wilken and Goggin 2015), and has been examined in relation to pioneering services like Dodgeball (Humphreys 2007, 2010, 2011), Brightkite and Loopt (Li and Chen 2010), through to Google Latitude (Ferrari and Mamei 2011; Page and Kobsa 2009), and Facebook (Abdesslem, Parris and Henderson 2010; C.-W. Chang and Chen 2014; H.-S. Kim 2016), Foursquare (Lindqvist et al. 2011), and Uber (Hjorth, Pink, and Horst 2018). Much of the existing scholarship on locative media has explored how end-users negotiate platform affordances of a range of platforms, offering nuanced accounts of "the fashioning of practices and devising of strategies for concealing, obscuring, or avoiding leaving behind locational traces" (Leszczynski 2015, 972). In the discussion to follow, I contribute to this literature by drawing on a study of Melbourne users of location-sensitive mobile social networking services (see the Introduction to Part II for further details), and how they negotiate and make sense of two quite different privacy-related issues: possible identification by unknown others while using these services, and data mining of these services for commercial gain within data markets.

Profile Pictures: Privacy as Identification

One of the fundamental structural tensions operative within many location-based social media services is that between "finding and being found" (Elmer 2010). People have used and continue to use these services, especially those that have a check-in facility, to find venues and to be found at these venues by those within their social network, and to find their friends as they check into the venues they have found. However, as discussed in Chapters 5 and 6, people tend to be quite particular in trying to manage how they disclose this information and to whom, and who can find them and under what circumstances. What this tension speaks to is an understanding (as noted earlier in this chapter) of privacy as concerned with *identification*. While concern over possible identification is something that is relevant to all forms of social media use, it is especially pertinent to and pressing for users of location-sensitive mobile social networking (LMSN) check-in services due to the potential for physical identification at specific times and places.

In order to get at this aspect of locational privacy, as part of my study of Melbourne locative media users I employed a "show-and-tell" method (Chamberlain and Lyons 2016; Sheridan and Chamberlain 2011), asking participants to walk me through the profile settings section of the Foursquare app on their phones (this was before the creation of Swarm app), discussing settings selection decisions and the reasons for their choice of profile picture. While identity construction is a recurrent theme within the available scholarship on social media account profile pictures (e.g., Hum et al. 2011; Kapidzic and Martins 2015), far less research focuses on the privacy implications and impacts of profile picture selection (see Debatin et al. 2009). What I found from discussions with my Melbourne participants regarding account settings was that the issue of privacy as identification emerged most strongly within discussion of profile pictures, and the reasons behind their selection.

Some of the Melbourne users I interviewed were quite happy to be potentially identified by other users of Foursquare as a result of their decision to reveal their faces in the profile pictures they selected for the profile section of the Foursquare app:

> I think the only thing that I made sure was that my bio was as similar as I can [make it] to my [other] public [accounts]. That's probably the thing I made sure [was] that, of all my public accounts, they're as similar as possible, so that I can [be] found. (Mario, market research, aged 35–44)

> It's just a picture I use that I'm happy with of my face. [. . .] It's easy to see. If someone runs to me in the street that has never met me before, but they've seen me on Foursquare, then they'll recognize me, but they'd have to be *really* familiar with the photo, which would be weird. (Elen, marketing, aged 25–34)

> I used to have a photo—like a cartoon. But, I just thought, if you check in [. . .] people want to know who you are, so I've got my photo now. (Mary, executive assistant, aged 25–34)

In Mario's case, his background in marketing led him to see benefits in being clearly identifiable and findable across platforms. Daniel Trottier (2018, 470) suggests that "individuals have various conceptual models of privacy which are balanced against other priorities"; in Mario's case, these priorities include selecting a profile image that ensures a unified social media presence that aids in findability and self-publicity. Elen was comfortable with the prospect of being found, but still considered the idea of someone recognizing her from her profile picture to be somewhat unsettling. Mary, when asked why she decided to change her profile from a cartoon image to an actual likeness, she responded, "I trust the settings on Foursquare." Mary attributed this trust (see Pink, Lanzeni, and Horst 2018) to the fact that, in her experience, settings changes—including opt-out options—were communicated clearly to her by Foursquare, which wasn't her experience of using Facebook. Adriana de Souza e Silva and Jordan Frith (2012, 128) note that "if users feel in control over their location information, these technologies are not perceived as a threat to personal privacy." This was the case for Mary, whose faith (whether founded or unfounded) in Foursquare's settings led her to feel confident that her privacy was protected.

Other interview participants, whose profile images depicted themselves, discussed how part of the reason they chose these images was because they considered them to be only partially identifying:

So, it's kind of semi-close up and I'm wearing sunglasses, so it's like you can kind of tell who I am but not totally tell who I am. It was in Greece, so it makes me happy when I see it, like being on the island in Greece, you know. Would that not make you happy? (Rebecca, marketing, aged 25–34)

Yeah, I have my own photo, which is, like, a very—it's a professional photo [in full make-up] that doesn't really look like me. So, I made sure of that. [. . .] Because, at the time, I still had 100 something friends [on Foursquare] and, I don't know, it was just something in the back of my mind [saying] don't put something [up] that can be recognizable straight away. It's true, I don't want to check into, say, here [the café in which the interview was conducted] and then some other person checks in and then he sees me, [sees] that I've checked in, you know; that was one of the reasons. Sometimes I use an avatar that's a cartoon, that I use on my blog, and things like that. (Caroline, executive assistant, aged 25–34).

I think [on Foursquare] it's a photo of a beer can or something. Oh, no, it's a photo of me, but it's not the same photo I've got on Facebook, on other social networks [. . .], so it's quite a—it's a bit [of an] obscurish photo, looking down on the face. It's a photo from having breakfast in Singapore actually, so this year. I don't remember putting that up there, but it's there now. (Marcus, media production, aged 25–34)

All three of the preceding passages speak to Alice Marwick and danah boyd's (2014, 1058) notion of "social steganography," which they describe as the practice of "hiding content in plain sight" through coded forms of identity performance. These same passages also provide rich examples in support of Woodrow Hartzog and Evan Selinger's (2013, 2015) suggestion that, when it comes to identification, "obscurity," as a "protective concept," might prove to be more productive than "privacy." The third of these passages is also striking for the way that it accords with Brittany Fiore-Gartland and Gina Neff's (2015) concept of "data valences"—the idea that there is often a discrepancy between what people say they do with their data, and what they actually do. For Marcus, it came as a surprise to learn that the profile image he thought he'd uploaded (a beer can) was not the one he had in fact uploaded (a semi-obscured image of himself that held personal significance as a marker or memento of a recent trip to Singapore).

Taking the idea of "obscurity" even further, some users preferred a profile image that was entirely unrelated to their own appearance. For instance, as Sophia (marketing, aged 25–34) explains, "I've got a picture of a famous internet cartoon on there as my profile." When asked whether that was a deliberate decision, she responded: "Yep. That way, if I check into a venue by accident and there's someone there who's, you know, stalking me [which, she revealed, had happened in the past], maybe they'll recognize the name but they won't be able to go, 'What's she look like? Where is she?'"

Finally, for Harry and Lucy, their profile pictures consisted of "joke images" (Gross and Acquisti 2005):

It's a picture of Slavoj Žižek. <laughs> I think when I picked that, I was just scrolling through the pictures on my phone and I thought, "oh, that would be funny." (Harry, student, aged 16–24)

It's a little—it's a baby girl on a skateboard. I was just going through my phone looking for a profile picture, and I was like, "Hey, that's really cute, I'll use it for that." (Lucy, student, aged 16–24)

While the use of "joke images" to convey humor and cuteness are clearly implicated in negotiations of online identity performance (Dale et al. 2017; Deumert 2014), they can also fulfill a protective function, with selection often made in order to preserve privacy by preventing possible personal identification (Gross and Acquisti 2005).

From these end-user accounts of profile images, there is a sense in which privacy is cast as risk mitigation (K. Burns 2013). In her book *Risk*, Deborah Lupton (2013, 153) suggests that "people tend not to have an overarching, all-pervasive fear of risk." Rather, "the risks they identify are highly discrete and contextual to their personal biographies" (153). What is more, they "are dealt with by most individuals at the level of the local, the private, the everyday and the intimate" (Tulloch and Lupton 2003, 7). This was very much the case for the participants I interviewed, especially with respect to how they negotiated the privacy risks they associated with the likelihood of being identified through their use of location service Foursquare. These personally negotiated privacy risks influenced the level of identification that participants were prepared to accept in the profile photographs they uploaded, but they also influenced their decision to enact other precautionary measures, such as employing "different layers of location to protect privacy" (Frith 2015b, 126; see also K. Burns 2013), like only checking into a suburb or an apartment building rather than an actual home address.

In the following subsection, I consider a further, crucial, privacy-related issue associated with mobile, location-sensitive social media services: the commercialization of end-user generated data.

Data Mining and Data Markets: Privacy as Data Security

Helen Kennedy, Dag Elgesem, and Cristina Miguel (2017, 271) argue that social media research rarely addresses directly the question of "what social media users think about the mining and monitoring of their social media activity." Moreover, when studies of attitudes to data mining are undertaken, the suggestion is that they tend to "fail to attend to diverse, individual and subjective responses to everyday tracking" (272). Social media data mining, as Kennedy, Elgesem, and Miguel (2017, 271) summarize it,

> includes a broad range of activities undertaken to analyse, organise, classify and make sense of such data, from counting likes and content shares to measuring reach, sentiment and key influencers, using techniques such as social network analysis, issue network analysis and natural language processing, among many others.

In the course of speaking with Melbourne locative media end-users, and in addition to asking them about identification-related privacy issues, I asked participants how they felt about the mining of end-user generated data and the monetization of these data through data markets (Elbaz 2012). In the responses that follow, I have sought to capture something of the "diverse, individual and subjective responses to everyday tracking" and related activities that were articulated by my respondents.

For a number of them, there was nothing exceptional about the fact that location-aware social media firms would capture and retain end-user data:

> I have no qualms about it, about organizations [like Foursquare] being able to use it [his data] from the back end. (Mario, market research, aged 35–44)

> Well, I'm in marketing so again it might skew my opinions, but I feel that, if you're willing to share that information [. . . with] the world, you have to expect it comes with a certain level of information availability. So, yeah, you put it out there, what do you expect? (Rebecca, marketing, aged 25–34)

> I think it's fine, if things are publically shared, that companies can data-mine if they want. (Harry, student, aged 16–24)

Caroline (executive assistant, aged 25–34) was also relaxed about the prospect of her social media data being mined. This was not just because her posts were public, but because she regarded what she posted to hold little commercial value (as she admitted to being quite cautious and vetted what she posted and when).

Others, however, expressed a desire for greater transparency on the part of social media corporations, and a clearer articulation of what they do with user-generated data:

> It'd be good if you are told, but are we ever told? I don't know, maybe they do [explain it] but I haven't read it. (Allen, architect, aged 55–64)

> I think it should be full disclosure. They probably are [disclosing what they're doing] for most of them, but it's probably hidden down the bottom somewhere in fine print and no one reads it. So, I've never actively gone out to find out what the policies are for these social media location-based companies. (Marcus, media production, aged 25–34)

Questions regarding data markets and the monetization of end-user data elicited a range of responses. For a number of respondents, there tended to be an understanding that the monetization of end-user data came with the territory: "I know that I'm signing up to the possibility of that" (Lucy, student, aged 16–24). Mario (marketing, aged 35–44) was accepting of what he saw as the trade-off involved in using a "free" service:

> I'm fairly open. If I was paying for Foursquare? Different deal. But, because it is free, I know that I'm the product. . . . So, if they sell [data] to a third party, then I have no issue with it because it's free. So, I know that that's the cost, the cost of having a free service. That's what it is. (Mario, marketing, aged 35–44)

Rebecca (marketing, aged 25–34) responded by noting terms and conditions (a.k.a. terms of service or terms of use) documents: "I mean, everyone has to click the 'yes, I agree with the terms and conditions,' right, when they sign up for anything like [Foursquare]?" While expressing doubt as to whether anyone actually reads these documents, Rebecca was of the opinion that

> as long as they're covering their legal bases [through terms of services documents] and informing where they should, then it's their [the platform's] prerogative and their right to work within the confines of the law. So, if we play within that constraint or those confines, then [. . .] that's how the world works.

What is clear from the preceding responses to the prospect of their own data being exploited commercially is that economic decisions are being weighed (K. Crawford 2012, 224) by these end-users. While this decision-making process has been given a variety of labels—ranging from "personal cost-benefit analysis" (Lehtiniemi and Kortesniemi 2017), "privacy calculus" (Dienlin and Metzger 2016; Krasnova, Trepte et al. 2017; Veltri and Günther 2012), and "privacy pragmatism" (Raynes-Goldie 2010)—all involve a weighing of the personal risks and benefits associated with use of these services.

Allen, however, held a slightly different view on the monetization of user data from that held by Lucy, Mario, and Rebecca. While Allen accepted that the monetization of data was to be expected, this did not prevent him from also expressing some degree of unease about this arrangement, evoking the boiling frog theory to describe how he felt we become less aware of what is being done with our data over time:

> I suppose it's okay [. . .] because you've agreed [. . .], so you can't use their stuff and say it's not okay. But, it's a bit odd, and, I suppose, maybe it's like the boiling frog theory: the water sort of heats up slowly around you, and you don't even notice that they're doing these things to you. (Allen, architect, aged 55–64)

The unease that Allen feels accords with Leszczynski's (2015, 977) notion of "anxieties of control," which she defines as "the impulse of wanting to *discern* (be informed of, voiced in terms of concerns with transparency) and *direct* (maintain the contextual integrity of) flows of personal locational information about oneself within and across networks, yet feeling that any attempt to do so is essentially futile."

Marcus, meanwhile, took a firmer, more direct stance, suggesting that, if end-users were the product, then there should be some form of compensation for the "immaterial" or "free" labor (Lazzarato 1996; Terranova 2000) that these end-users perform for platforms:

> I suppose I'm angry I don't get anything out of it. There's people out there in my mind [who] are getting rich off this stuff [. . .] and I'm getting nothing for it, so maybe give something back to the users and that might make it okay; that'd make it okay for me. (Marcus, media production, aged 25–34)

While he saw little evidence of it, Marcus regarded "fairness" to be an important aspect of the transactional nature of online labor—a position that accords with Kennedy, Elgesem, and Miguel's (2017) findings from their study of social media users' attitudes toward data mining.

In the preceding section, I have presented the views of a number of Melbourne-based users of mobile locative social media services. These insights add to a growing body of work that explores the manifold ways that end-users of location-sensitive mobile social networking "articulate and perform privacy in the wake of significant legal, technical and market-based pressures" (Trottier 2018, 473; see also boyd and Marwick 2011; Leszczynski 2015).

While it remains vitally important to continue to seek to understand end-user privacy practices and privacy perspectives, there is, however, growing critical understanding that individual measures do little to counter the increasingly ubiquitous and voluminous capture of personal data by social media corporations and other agents. As Leszczynski (2015, 966) puts it, "personal locational disclosures cannot be directly or easily controlled by individuals through adjustments to settings on any one device or within any one application or service." The rise of "big spatial data" in particular presents manifold difficulties for the forms of location privacy self-management (Solove 2013), "privacy pragmatism" (Raynes-Goldie 2010) and "consent dilemmas" (Solove 2013) that have been documented in the preceding discussions, effectively rendering them "practically unfeasible" (Leszczynski 2015, 977). In the sections to follow, I explore in more detail why it is that individual measures to protect personal privacy are viewed increasingly as largely ineffectual, and then I canvass a range of scholarly and emerging regulatory responses to the capacious capture and commercialization of personal (geocoded) data.

THE PRIVACY CHALLENGES OF "SPATIAL BIG DATA"

In the wake of the rise of the networked society (Johns 2017; Rainie and Wellman 2012) and the networked economy (Shapiro and Varian 1999), we have, as Helen Nissenbaum (2011, 33) has suggested, "witnessed radical perturbations in the flow of personal information" and a significant scaling up of the mass extraction, collection, and trading of personal data, especially personal spatial data. As Agnieszka Leszczynski (2017, 236) notes, "our personal spatial data [now] flow freely and without friction across and between interoperable and synergistic geo-enabled devices, services, applications, and analytics engines." The scale of these data-capture efforts and the motility of these data (Coté 2014) render largely ineffectual the dominant "notice-and-consent" approach to managing and addressing privacy concerns, individual attempts to access the data that is generated by and about them, as well as individual data and privacy self-management efforts. Daniel Solove (2008) refers to these three impacts as problems of transparency, exclusion, and aggregation; each warrants some further explanation.

Transparency

Helen Nissenbaum (2011, 34) has argued that transparency and choice, or notice-and-consent, became the dominant approach to managing privacy concerns because of "the popular definition of a right to privacy as a right to control information about oneself," and because transparency and choice mechanisms take a relatively light regulatory approach that appeals to a competitive free market. However, notice-and-consent processes have been roundly critiqued by both legal and communication scholars (Andrejevic 2010; Fernback and Papacharissi 2007; Livingston 2011; Nissenbaum 2011; Solove 2008). Terms and conditions and notice-and-consent documents fail, Nissenbaum (2011) argues, for a range of reasons, including the ease with which they can be altered at will by the corporations that issue them (35) (and, because of this, are subject to "scope creep"), and as a result of what she dubs the "transparency paradox," whereby "transparency of textual meaning and transparency of practice conflict in all but rare instances" (36). On this second point, Daniel Solove (2013, 1882) notes that "nearly all instantiations of FIPPs [Fair Information Practice Principles, which include terms and conditions documents] fail to specify what data may be collected or how it may be used" (see also Andrejevic 2010, 87), especially location data (de Souza e Silva and Frith 2012, 128). Thus, "private," in the context of terms and conditions documents, "pits platform owners' accumulation of personal data against users' expectations of self-determination in the outcomes of that data" (Trottier 2018, 473; cf. Elmer 2015).

What should also be acknowledged here is that there is growing recognition that the push for greater transparency—which has often been viewed as an ideal—comes with its own limitations. Not only does transparency not always achieve its desired outcomes, especially in terms of generating greater understanding of algorithmic systems (Ananny and Crawford 2018), but following the stipulation for greater transparency to the letter poses a range of challenges, from potentially increasing the cost of information, to proving overly burdensome for consumers and possibly dissuading them from better informing themselves of data-collection practices (Rhoen 2016, 5; McDonald and Cranor 2008).

Exclusion

Writing on the cusp of the explosion in popularity of social networking, Daniel Solove (2004, 13) notes the rise of what he terms "digital dossiers." "We currently live in a world where extensive dossiers exist about each one of us," he writes. "These dossiers are in digital format, stored in massive computer databases by a host of government agencies and private-sector companies"—including, increasingly, social media firms. What is more, these dossiers are increasingly automated—populated by the accrual of "large volumes of information captured by a distributed array of sensing devices" (Andrejevic and Burdon 2015, 27), and now contain more data *about* users than *by* users of location-sensitive mobile devices. Thus, these digital dossiers and the data contained in them exist at an increasing remove from the individuals featured in them, where data-processing operations "are for the most part invisible, managed at distant centers, from behind the scenes, by unnamed powers" (Obar 2015, 1). One result, Jordan Frith (2015b, 122) suggests, is that, even if people are not actively surveilled upon, "their privacy is still impacted by the massive collection of information because they cannot find out what information different entities possess" (see also de Souza e Silva and Frith 2012, 128–129).

This increasing disconnect between data producers and data aggregators, along with a lack of knowledge about data traces, Obar (2015, 1) argues, produces a "perpetual information illiteracy—an intellectual detachment from the rapidly expanding universe of Big Data." Information illiteracy is also further exacerbated by corporate and legal attempts to stymie or block access that might help develop consumer knowledge about data traces. In a notable Australian legal case, Australia's Federal Court ruled in favor of telecommunications provider Telstra, in a case brought against it by the Australian Privacy Commissioner as to whether or not "telcos" should have to provide stored metadata (which can include, among other information, IP addresses, URLs, and specific cell tower location information) to its customers upon request (Biggs 2017). The case was initially prompted by technology journalist Ben Grubb's failed requests that Telstra provide him with a copy of his full metadata. Telstra would only supply him his billing information (Biggs 2017). While reports of the outcome of this case suggested that the Federal Court ruling meant that Telstra was not required to fill Grubb's request, and is not required to fulfill future requests for full customer access to full metadata (Biggs 2017), the result was in fact much less clear-cut than this and remains open to legal debate. As Anna Johnston (2017, n.p.) writes,

> The court made *no decision* about whether or not the metadata was "about" Ben Grubb, because it wasn't asked to. The court made *no decision* about whether or not Ben Grubb's identity could be ascertained from the metadata (alone or in conjunction with other data), because it wasn't asked to. The court made *no decision* about whether or not Ben Grubb's metadata was "personal information,"

because it wasn't asked to. This case was about a question of law, not the application of that law to a particular set of facts. The only thing decided [. . .] was that the phrase "about an individual" is an important element in the definition of personal information, *as the definition existed in 2013.*

The contested nature of the legal outcome notwithstanding, the protracted nature of this case (2013–2017), the willingness of Telstra to pursue it, and the miring of the case in "a question of law" underscore the difficulties consumers have faced and continue to face in being able to access personally identifiable data and data traces that are increasingly about them.

Writing in 2012, Adriana de Souza e Silva and Jordan Frith suggest that the broader issue here is that, "as companies collect more and more data to build increasingly robust profiles, people have little recourse to access what information has been collected or whether that information is correct" (2012, 128–129), and consequently "have little control over what is done with their own locational information." However, this has shifted with the introduction within the European Union, from May 25, 2018, of the General Data Protection Regulation (GDPR), which has meant that people in the European Union *can* now ask for copies of their data; social media firms, such as Facebook, Instagram, and Twitter, also now allow consumers to obtain copies of their data, although questions remain about the scope and detail of the data these logs contain. Even so, for end-users of location-sensitive social media services, the more general problem of personal data management remains unchanged: "there are too many entities, too many quickly moving parts, too many stockpiles and too many data points to expect a consistent, exhaustive and ubiquitous data privacy self-management" (Obar 2015, 10; see also Esayas 2017).

Aggregation

On the issue of aggregation, Daniel Trottier (2018, 470) suggests that privacy protections that imply a degree of control over personal information flows have been "greatly complicated" by social and search media and the rise of big spatial data. The sheer scale of the data aggregation efforts of these firms, combined with cross-platform deals, data-sharing arrangements, and the "data motility" within and between firms (Coté 2014), pose significant challenges to end-user understanding of data retention and individual privacy. A key focus in existing scholarship on locative media and privacy has been on emphasizing that users' negotiations of locational privacy is, and ought to be, "intimately related to the ability to control the context in which one shares locational information" (de Souza e Silva and Frith 2012, 129). To use Foursquare as an example, given the multitude of sources that feed Foursquare's places database, the fact is that "people who use a variety of location-based services are interacting with Foursquare data without even knowing it" (Frith 2015b, 105). Moreover, even if end-users think they are clear about their own privacy settings, the entangled nature of the app ecology and their own (partial) knowledge of corporate arrangements, algorithmic processes, terms of service documents, and device settings complicate matters considerably. This is a point that has been made forcefully by a number of critics. For example, Samson Esayas draws from a detailed examination of the data-combination practices of Google and Facebook to make the point that "the focus on individual processing activity overlooks the fact that the totality of personal data collected" is often drawn from sources that originally served distinct purposes, and/or is the result of "the combination of data across these processing activities" (Esayas 2017, 144; see also Cuijpers and Pekárek 2011). In addition, as Kate Crawford (2012, 224) observes, "the serious problems facing the rapidly expanding field of locative data [. . . are] the nature of the bargains being struck, and whether all parties have full knowledge of what role they are playing and where it ends."

RESPONSES TO THE PRIVACY CHALLENGES OF "SPATIAL BIG DATA"

In the final section of this chapter, I canvass a number of responses by legal, communication, and other scholars to the aforementioned privacy challenges of "spatial big data." I start by returning to Helen Nissenbaum's privacy as contextual integrity framework.

Nissenbaum's work is significant in this context in that it arguably remains the most nuanced theory of privacy that attempts to accommodate and respond to the sorts of privacy challenges presented in the preceding section. To recap: the contextual integrity framework "characterizes privacy as appropriate information flow, and appropriate flow characterized in terms of three parameters: actors (subject, sender, recipient), information type, and transmission principles" (Benthall, Gürses, and Nissenbaum 2017, 3). While Nissenbaum (2010, 216) acknowledges that "the prodigious capacity of information technology and digital networks to publish and disseminate information has given rise to a host of privacy problems and puzzles," she is confident that the privacy theory challenges presented by these technological developments are not insurmountable. The privacy as contextual integrity framework, she argues, provides a critical lens or "heuristic" that can be applied at different scales and across a range of contexts, and is robust enough to address the aforementioned privacy problems and puzzles. She writes:

> The challenge of privacy online is not that the venue is distinct and different, or that privacy requirements are distinct and different, but that mediation by the Net leads to disruptions in the capture, analysis, and dissemination of information as we act, interact, and transact online. The decision heuristic derived from the theory of contextual integrity suggests that we locate contexts, explicate entrenched informational norms, identify disruptive flows, and evaluate these flows against norms based on general ethical and political principles as well as context-specific purposes and values. (Nissenbaum 2011, 38)

In practical terms, the challenge of privacy online is addressed by "look[ing] for the contours of familiar social activities and structures" and, once these are identified, "bring[ing] into view the relevant norms" (43). To illustrate and give further substance to her arguments, Nissenbaum applies her contextual integrity heuristic to a number of cases that are germane to the concerns of this chapter and the book as a whole: Google Street View; RFID tags; online search; data aggregation and analysis; and, the implications company mergers have for data management and privacy (Nissenbaum 2010, 2011).

Despite the undoubted impact and continued importance of the contextual integrity framework for thinking through contemporary privacy-related challenges, Nissenbaum's model has not been without criticism. In reading *Privacy in Context*, it was not always clear to me who she imagined would be responsible for applying the contextual integrity heuristic, and who it is who gets to determine appropriate norms for each context or situation. Is it legal scholars, or legal representatives and the judiciary? Is it industry? Is it regulators and policy-makers? Is it public citizens? Is it all of the preceding? These are important questions to be asking, particularly as it strikes me as difficult to grasp how end-users of search and social media and other related services are likely to always have the capacity or skills or inclination to weigh up, on the fly, possible privacy breaches against normative legal expectations.

In addition, in his review of *Privacy in Context*, Rocco Bellanova (2011, 393–394) makes an important basic observation: Nissenbaum's model is itself contextually specific in that it has been developed within a US legal context, and, as such, requires further "evaluation" as to its suitability to other legal jurisdictions and contexts, such as the European Union and elsewhere.

More problematic for Bellanova are the implications of Nissenbaum's model for what he terms a "politics of privacy." One of the explicit aims of Nissenbaum's model is to "sift legitimate claims to privacy from non-legitimate ones" (Bellanova 2011, 394). However, this sifting process comes at a cost. For Bellanova, it evacuates "the mapping of the actual political uses of privacy (and data protection), if not [. . .] most of politics":

> Think about the use of privacy or data protection for contesting the integrity of an established context from within, for example when personal data are already processed by an existing socio-technical system deemed to be legitimate by the lack of public debate or even by the consensus of the majority. Or, on a quite opposite note, think about the discursive rhetoric of many EU-US agreements claiming to protect and enforce data protection, but *de facto* establishing new security systems. (Bellanova 2011, 394)

Here, Bellanova asserts, privacy is relegated to a "sort of 'political vacuum,' where 'contexts' appear politics-free, empty of constitutive struggles or power relations" (394), and where "only contexts, and related values and rules [and not politically-engaged or motivated actors] seem to count" (395).

For Gabe Maldoff and Omer Tene (2017), Nissenbaum's contextual integrity framework—especially as it has been applied to the NSA's bulk telephony metadata collection in the United States (Kift and Nissenbaum 2017)—makes clear that a contextual violation did occur. Maldoff and Tene's (2017, 396) contention, however, is that, when it comes to big data, "the question is not whether there is a contextual violation," but rather, whether and when such a violation is or might be justified. What is required, they argue, is a "second step"—performing a cost-benefit analysis—and that, at present, the contextual integrity framework "offers little guidance on how to tally data [retrieval and collection] benefits" (396).

In addition, critics such as Agnieszka Leszczynski and Mark Andrejevic suggest that Nissenbaum has significantly underestimated the scale and complexity of data markets and associated data transfer impacts and implications. As Leszczynski (2017, 236) notes, "Our personal spatial data flow freely and without friction across and between interoperable and synergistic geo-enabled devices, services, applications, and analytics engines." What is more, the collection and disclosure of an individual's locational data are *uniquely pervasive* (continuous), *platform independent* (collection and disclosure of locational information are not particular to any one application, but is rather implicated in digital practices across an array of mobile services, nor is it specific to any one device or mobile operating system), and *indiscriminate* (implicates anyone who owns a mobile device rather than being dependent on voluntary participation on designated platforms). (Leszczynski 2015, 966)

In light of this, it thus becomes increasingly difficult to assess contextual violations and determine normative expectations. Furthermore, as Andrejevic and Burdon (2015, 32) point out, the big spatial data analytic process is "systemically and structurally opaque," and "it follows that data collection and analytical infrastructures are equally opaque." Because of this,

processes of opacity that yield un-anticipated uses for data that result in uninterpretable decisions undermine some of the key foundations of information privacy law, namely, informed consent and even ideas such as contextual integrity. (Andrejevic and Burdon 2015, 32)

The wording of this passage is significant insofar as Andrejevic and Burdon are acknowledging both the importance of Nissenbaum's model of privacy and the fact that no single privacy model is sufficient given that we are now dealing with data-collection processes operating at such scale and rapidity that we struggle to comprehend the possible end-uses of these data (see also Solove 2013, 1881)—a situation that also partially explains why it is that privacy is often entangled with concerns that are increasingly inseparable from it, like data security. Indeed, the exponential increase in the collection, aggregation, and automated analysis of big spatial data—combined with increasingly frequent reports of questionable data-handling practices, concern over the influence of data brokers (Crain 2018), and data security breaches (such as the leaking of sensitive personal data, including HIV status, of Grindr users)—has contributed to new waves of concern and a wide range of approaches that range from the discursive, to the regulatory and policy-related, to the legal.

At the discursive level, there have been many and varied responses. Critical GIS scholar Agnieszka Leszczynski (2017), for instance, calls for further conceptual reinvigoration and reorientation of privacy theory that gives specific emphasis to the circulation, capture, storage, and commercial uses of spatial data—what she terms "geoprivacy." For Leszczynski (2017, 236), "geoprivacy" encompasses more than just location. It must "account for the emergent complex of potential privacy harms and violations that may arise from a number of nascent realities of living in a (spatial) big data present." These emergent realities, as Leszczynski sees them, are fourfold at least:

(1) from the spatial-media-enabled pervasive capture and repurposing of individuals' personal spatial-relational and spatio-temporal data;
(2) from the ways in which individuals cast digital footprints as they move across numerous sensor networks of smart cities;
(3) from the circulation and analytics of these data, which position individuals as spatially precarious in various and unprecedented ways; and
(4) from the inability of individuals to control highly personal flows of spatial information about themselves in networked device and data ecologies. (Leszczynski 2017, 237)

For Leszczynski (2017, 242), what is significant about this understanding is that it departs from "highly axiomatic definitions that stress the individual and a negative definition of privacy rights" to propose instead a conception of privacy that "encompasses and accounts for the realities of continuous personal locational data flow as a feature of everyday digital practices that are characterised by extensive, real-time geosurveillance and the networked data and device ecologies of spatial media." What this understanding of "geoprivacy" is yet to provide, however, is a detailed account of its socio-legal application, and how this might build on and/or depart from Nissenbaum's contextual integrity framework.

Greg Elmer (2015), however, questions whether a shift in focus from privacy to *publicity* might be warranted. While acknowledging the ongoing importance of privacy concerns, Elmer wonders "how long privacy advocates can continue to shine [a] light on the individual, when social media platforms always and already profile clusters of users on or off platform, with or without ever consenting to a terms of service" (Elmer 2015, 1). He asks:

> Would it not make more sense, given the stated goals of social media CEOs, the intended goals of social media platforms, and the process of public stock listing, to develop critical theories of publicity, as financialized on social media platforms and beyond? (Elmer 2015, 1)

While not precisely focused on questions of publicity, critical theory of the sort that Elmer is calling for—that is, work that moves beyond the rights of the individual to consider platform profiling of clusters of users—is already emerging, as evidenced in scholarship on the "like economy" (Gerlitz and Helmond 2013), the "attention economy" (Williams 2018), and geodemographic and algorithmic sorting (see, for example: Barreneche 2012a, 2012b; Beer 2016; Brodmerkel and Carah 2016; Bucher 2012a, 2012b, 2013b, 2017; Carah 2014; Esayas 2017; L. Evans 2014; Noble 2018; Pasquale 2015; H. Smith 2017).

Meanwhile, legal scholar Angela Daly and colleagues have taken a quite different tack, embarking on a project—@good_data—that seeks to mount a constructive response to the sorts of privacy and data security concerns raised in this chapter by exploring and establishing what "good data" and "ethical data practices" might look like, "with a view to developing policy recommendations and software design standards for programs and services that embody good data practices, in order to start conceptualising and implementing a more positive and ethical vision of the digital society and economy" (Carlson 2018; see also Daly, Devitt, and Mann 2019).

In addition to the preceding discursive responses, a key socio-technico-legal response to the limitations of notice-and-consent systems, and the broader privacy impacts of spatial, social, and search media, has been that of "privacy-by-design"—an approach where privacy and data protection are promoted from the start and built in (rather than added on to) projects and technologies. While this approach promised much, numerous studies have since revealed that the "privacy-hardwiring enterprise"—as Lee Bygrave (2017, 756) refers to privacy-by-design—has largely failed to live up to its potential (see Bygrave 2017; Bellanova 2017; Koops and Leenes 2014; Rubinstein and Good 2013). This has been for a host of reasons, including, as Bygrave (2017, 756) argues, because "such an enterprise [. . .] is at odds with powerful business and state interests, and simultaneously remains peripheral to the concerns of most consumers and engineers"; as legal scholar Julie Cohen (2012, 110) puts it, "in the networked information society, protection for privacy compromises the liberal commitments to free flows of information."

Another "imperfect yet pragmatic" suggested socio-technical response to the difficulties presented by data privacy self-management is that of "representative data management" (Obar 2015, 13). What this involves are intermediaries or brokers working in individual consumer interests to develop systems and services for the responsible collection and management of their (spatial) big data (Obar 2015).

In terms of policy-related responses, in Australia there have also been calls for the introduction of digital rights that address consumer concerns about privacy violations by corporations, the use of data analytics and targeted advertising, the centralization of data by government and other actors, and a desire for greater regulation of online discussion environments and social media platforms (Goggin et al. 2017).

And, finally, as the writing of this book was coming to an end, the most significant legal response to date was introduced in the European Union: the General Data Protection Regulation (GDPR). The GDPR regulation replaces the long-outdated data protection directive from 1995 and came into force in May 2018. It gives individuals considerable more power over their personal data, both for "transactions that occur within EU member states and over the exportation of personal data outside the EU" (Chakravarty 2018). The implications of the GDPR for location data are significant in that, under Article 4 (1) of the GDPR, location data are considered "personal data," and "personal data are granted extended rights, including a right to access and a right to erasure" (Chakravarty 2018). Given these requirements, it is interesting to note that Facebook moved quickly to relocate its 1.5 billion people user base in Africa, Asia, Australia, and Latin America from Facebook Ireland, which is subject to the GDPR regulation, to Facebook Inc. in the United States, which is not (or at least not to anywhere near the same extent) (Chaturvedi 2018).

CONCLUSION

This chapter has explored the vexed issue of privacy in relation to mobile location-sensitive social media services. This exploration of the issue of privacy proceeded in three steps.

The chapter began by exploring a number of significant conceptions of privacy that have proven influential within social and locative media scholarship, and by detailing why it is that geographers and media scholars maintain that the geolocation data gathering and sharing capacities of internet-enabled smartphones have brought additional layers of complexity to our understandings of the privacy implications and impacts of our use of social media apps and mobile devices.

In the second section of the chapter I explored (a) how these privacy implications and impacts—particularly as they relate to the issue of personal identification—have been negotiated by Australian consumer end-users of these services, and (b) sought their opinions on the extraction and commercialization of user data. With respect to the first of these, while some participants were happy to be identified through their use of location-sensitive social media services, most employed a variety of strategies to manage and generally lessen this possibility. With respect to their views on data markets, while those working in marketing and parallel industries were comfortable with the commercial exploitation of end-user data, others expressed a degree of unease—"anxieties of control" (Leszczynski 2015)—about these arrangements, but a sense of powerlessness to change them. There are of course many further avenues to explore that can further extend this analysis of everyday privacy practices and attitudes to privacy, including (but by no means limited to) how these issues are played out within developer communities (Greene and Shilton 2018), within business contexts, both small and large (Humphreys and Wilken 2015; McCosker 2017c), cross-culturally (Humphreys, Pape, and Karnowski 2013), and within vulnerable or marginalized groups (Marwick and boyd 2018); exploring these issues in and across such contexts was, however, beyond the scope of this book.

Finally, in the third section of the chapter I turned to examine the challenges presented by the rise of "spatial big data." While earlier in the chapter I suggested that location has often served as a lightning rod for debate around technology use and privacy, the privacy challenges before us are now of a vastly different level of complexity. This is not to say that location and geodata no longer matter or are no longer implicated in data-capture processes—far from it. As Mapbox's Hannah Judge puts it, "every data set has a spatial component" (Brassey 2018, n.p.), and location data—as this book has argued at length—remain extremely valuable and highly sought after (Chakravarty 2018). What has shifted is that location data are now integrated into a wide array of services that are data-mined with increasing voraciousness and at a scale that arguably renders established models of managing privacy inadequate. What is more, with each new revelation about questionable data-handling practices, security breaches, and data leaks comes growing consumer consternation and mounting pressure on corporations, policy-makers, regulators, and legislators to respond. Whether or not the introduction of GDPR regulations in the European Union heralds a significant wider global shift in momentum from the sorts of data-mining freedoms that search and social corporations have enjoyed to date and toward greater regulation and legal strictures remains to be seen. What does seem clear, however, is that we are (at time of writing) standing at a particularly fascinating critical, cultural, and legal juncture.

Conclusion

Location [is] at the heart of technology shifts.
—Chris Sheldrick, cofounder and CEO, What3Words (quoted in Geoawesomeness 2016)

This book has sought to draw out the manifold ways in which location, location awareness, and location data have all become familiar yet increasingly significant parts of our mobile-mediated experiences of everyday life. Adopting a cultural economy framework, the book has explored the complex of interrelationships that mutually define cultural usages of locative media, and the new business models and economic factors that emerge around and structure locative media services and their diverse social uses and cultures of consumption. By adopting this approach, the book has sought to develop a coherent, systematic, and in-depth account of how location-based services, such as GPS-enabled mobile smartphones and associated applications, are socially, culturally, economically, and politically produced and shaped, as much as technically designed and manufactured.

In pursuit of these concerns, the book's contents were divided into three parts, each with its own specific focus, but that also connected with the concerns of the other parts and chapters. In the Introduction, I suggested one way to think of these parts: as separate rooms, each exploring its own specific curatorial concerns, that combine to form a larger exhibit. An alternative way of thinking about the book's structure is to draw a comparison with a kaleidoscope, with each part (and the chapters it contains) representing a turn of the kaleidoscope that brings into view a particular arrangement of themes, issues, questions, and actors. In combination, the book as a whole offers a rich, composite portrait of locative media in its cultural economic complexity.

Part I, titled "The Topography of Location Media," examined the industrial composition of mobile location services, and traced the race to control digital and mobile maps, and the rise and ongoing evolution of location-based mobile social networking and search and recommendation services. Chapter 1 took an ecosystems approach to examining mobile location-based services, and the larger media ecologies these services intersect with and depend upon. The chapter sought to give form and shape to the field of mobile location-based services, and, in so doing, explore *what* these services are, *who* the key corporate players are, and *how* they interact. Chapter 2 examined the political economy of digital mobile maps, and charted the ambitions and struggles of Apple as it set about building a digital maps service to rival Google's. From this examination, a clearer picture emerged of the very significant technological and financial investments required to build and maintain high-quality maps, and the uncertain and still emerging business and revenue models associated with these maps-related efforts. Chapter 3, the last chapter in Part I, examined the geolocation data capture efforts of two firms now seen as central to the contemporary settlement of locative media: Foursquare and Facebook. The chapter sought to develop a clearer picture of how the integration of location data has changed over time, and how the means of extracting economic value from these data have matured in tandem with the growth of social media platforms, the embrace of datafication, and the emergence of data markets. For both these firms, the locational traces of our passage are highly prized and financially lucrative.

Part II, titled "Cultures of Use," opened with an examination of key locative media arts projects that creatively engaged with mobile phones and their supporting infrastructures to explore the socio- and geo-political possibilities and implications of these locative technologies, and how art could serve as criticism, as an enactment of a subtle political aesthetics. In taking up these themes, Chapter 4 explored three projects—Blast Theory's *You Get Me*, Josh Begley's Metadata+, and Julian Oliver's *Border Bumping*—and how each, in its own way, created the conditions of possibility (a "pedagogical invitation") for thinking through aesthetics and its relationship to politics. The focus of Part II then shifted to an examination across two chapters of how location-sensitive mobile social networking services, and people's uses of them, form part of complex communicative ecosystems that are intimately involved in the negotiation of everyday life, and of subjectivity and identity formation. Chapter 5 advanced the claim that there is a tendency toward classification that involves forms of compartmentalization in the use of location-enabled mobile social networking services that shape app selection and use. What emerged, I argued, was a rich and complicated picture, where choices governing the uses of, and shifts between, various applications could be seen as the result of finely granulated individual and social practices around the management of aspects of one's social networks and app use, and as a result of shaping by socio-technical dynamics, such as the specific "affordances" and constraints of each application or platform. Understanding the complexities driving these end-user choices of application selection and interaction, I argued, remains crucial if we are to more fully grasp the ties that bind platforms, political economies, and publics, and if we are to critically respond to key policy considerations, such as the privacy impacts and implications of location-based services. Chapter 6 examined how urban spaces and places are explored, catalogued, and communicated, and how these communicative practices are entwined with individual identity negotiation, performance, and display, through the use of location-sensitive mobile social networking and search services. What shone through from this examination was the need to take seriously, and continue to further think through, the complicated identity work that is undertaken by users through locative social media and search and recommendation platforms *and* as a result of their own self-enrollment in commercially focused contexts of use.

Part III, "Geodata Capture and Privacy," the final part of the book, examined how and why it was that the gathering and use of geocoded data became of such concern to private corporations, government, state actors, consumer advocacy groups, and individual consumers. Chapter 7 explored two key and controversial cases—the Google Street View Wi-Fi data collection scandal, and the geolocation information capture efforts of the NSA as disclosed through the Snowden documents—where end-user geolocation data have been accessed, accrued, and stored outside of end-user knowledge and consent, and which push up against the boundaries of accepted regulatory and legal frameworks. Both of these cases changed significantly the tenor of debate around state-corporate trust, data protection, and privacy. Finally, Chapter 8 examined the vexed issue of privacy in relation to mobile location-sensitive social media services from a range of perspectives. It explored fraught definitional understanding of the concept of privacy, how geolocation data-gathering and sharing capacities of internet-enabled smartphones have brought additional layers of complexity to our understandings of privacy, what end-users do in negotiating privacy through everyday practice, and how the complexities of "spatial big data" pose new and even more significant privacy challenges. With respect to the last of these, what has shifted with the rise of spatial big data is that location information is now integrated into a wide array of services that are data-mined with increasing voraciousness and at a scale and speed that render established models of conceptualizing and of managing privacy increasingly inadequate.

The aim, in employing different analytical lenses or foci (different turns of the kaleidoscope) across the three parts of the book—industrial composition and business and revenue models; consumption and personal and social use; and policy and regulatory concerns and contexts—was to develop a rigorous, comprehensive, and coherent, interdisciplinary account of location-based technologies and locative media and their social and cultural economic impact and significance.

* * *

Technology, it has been said, rarely stands still (Ziman 1980, 76).

On one of Apple's developer sites, details are provided as to how location determination is managed in iOS, through what they call their Significant-Change Location Service:

> The significant-change location service offers a more power-friendly alternative for apps that need location data but do not need frequent updates or the precision of GPS. The service relies on lower-power alternatives (such as Wi-Fi and cellular information) to determine the user's location. It then delivers location updates to your app only when the user's position changes by a significant amount, such as 500 meters or more. (Apple 2018)

Apple's Significant-Change Location Service provides a useful metaphor for explaining the shifts that occur within the field of locative media and location-based services: certain things take prominence, then a shift—a significant change—occurs within the field, and other things take prominence, and so it continues. Many of the forms and uses of locative media that were emergent at the outset of this project have waxed and waned over the life of this project, with some remaining in use (Swarm), some long since gone (Whrrl, Brightkite, Gowalla, Loopt, Sonar), and still others in slow decline, or having transmogrified almost beyond recognition (Foursquare, Banjo); in addition, other, newer forms and uses have since emerged, or are just beginning to emerge.

At one point during a Skype interview with New York–based entrepreneur and angel investor Fabrice Grinda, he declared that "the only thing inherently limited around location-based services is that they're location-based" (pers. comm., 2013). I understood this quip at the time to be a playful provocation and prompt to look and think beyond the "check-in" service and of location as something one actively registers via a smartphone. And there is sense in this. For, while the "check-in" as a specific function of mobile use has lost its novelty—as happens with all new media forms, whether current or past (Farman 2015c)—and while location-based mobile social networking services appear to have fallen out of favor, it remains the case that "their architecture and functioning are becoming stable parts of other, bigger social networks" (L. Evans and Saker 2017, 96) and a "normal, integrated aspect of social media use" (95).

Normalization of location-awareness should not, however, be mistaken for declining interest in the possibilities of location sensitivity, and how these might be developed in new and novel ways. As Brightkite's co-founder, Martin May put it, "Location is really just a feature. It makes sense in a ton of different apps. It makes sense for car hailing, for all the maps applications, navigation, and social interactions—there's a ton of different applications for location" (Martin May, pers. comm., 2013). Indeed, location determination and location data-extraction systems are now being developed and deployed in diverse and at times unexpected ways that are yet to be fully explored and understood.

Location-related technologies are, for instance, central to contemporary visions of and growing investment in "smart" or "networked" cities (Kitchin et al. 2019; McQuire 2016; Townsend 2013; Coyne 2010). Location technologies are also increasingly being coupled with vision capture technologies. In a 2017 *New York Times* article, Brian Chen (2017, n.p.) suggests that the future of the smartphone will be "all about the camera." The basis for this claim is the integration of depth-sensing vision capture technology into next-generation iPhones, Android devices, and other systems. This technology is powered by chips, developed by the likes of Qualcomm and Movidius, that permit phone cameras to scan and "map distance and 3D contours or an object or location, with precision to a fraction of a millimeter" (Captain 2017). For instance, at its I/O 2017 developer conference, Google announced its "Visual Position Service," a refinement of their long-standing experimental "Project Tango." Google's "Visual Position Service" has been described as "a collaboration with Google Maps where indoor environments are mapped out and tagged within the view of a mobile phone camera" (Mueller 2017). In addition, Google has also announced a smartphone image-recognition feature, called Lens, that permits the phone camera to recognize complicated images and to identify and act on information contained within these images, such as providing full business, location, and other information of a restaurant when the camera is aimed at shop signage from across a street (Perez 2017). Depth-sensing vision capture technology of the sort noted here is likely to have wide application, including in areas such as AR (augmented reality) and VR (virtual reality), with Google releasing its augmented reality feature for Google Maps (Heater 2018), and in drone vision (McCosker 2015a, 2015b) and remote sensing (Ambrosia, Hutt and Lulla 2011), and elsewhere.

Location-centered depth-sensing vision capture, precision mapping, as well as an array of other sensing technologies, are also all of vital importance to autonomous vehicle development (Etherington 2018; Mlot 2015; Quain 2017). Precision maps are vital at the orientation and decision stages of autonomous vehicle operation (Wilken and Thomas 2019; Alvarez León 2019); as one article puts it, "What HD maps give self-driving cars is the ability to anticipate turns and junctions far beyond sensors' horizons" (Economist 2016). The particular challenge for driverless cars, as autonomous vehicle development pioneer Sebastian Thrun explains, is to "map the environment while simultaneously determining the [car's] position relative to this map" (Thrun and Leonard 2008, 872). "Key enablers" in responding to this challenge are what Thrun and Leonard refer to as "simultaneous localization and mapping" (SLAM) processes (871), whereby the vehicle operates as a communication platform to combine, in real-time, finely granulated precision maps information to orient the vehicle, with supplementary perception data generated from the vehicle's arrays of sensors, and with both sets of data continually interpreted and acted upon, then reinterpreted and acted upon, and so on, by the decision-making central processing unit (CPU).

Location-sensitivity has also come to form a crucial component of the "lively data" (Lupton 2016, 4) captured by "smart watches" and other self-tracking devices (Neff and Nafus 2016) that are part-and-parcel of our contemporary fascination with "miniaturized mobilities" (Elliott 2016, 168–178). Precise location data generated by, and through, our ongoing obsession with portable, connected devices are viewed as crucial to the planning of "smart cities of the future" (Cunningham 2017), the successful operation of the internet of things (see Bunz and Meikle 2018, 26–44), and are also being used to track and interpret indoor foot traffic flows (via the use of Apple's iBeacons, for example) and in driving the development of indoor maps. In the case of indoor maps, precise positioning information is accessed by a variety of means, including from phone sensor data, radio fingerprinting, inertial navigation, variations of the geomagnetic field, and so on. In an interview with Sina Khanifar, cofounder of OpenSignal, a service that measures mobile telecommunications signal strength, he discussed his team's branching out into smartphone sensor-generated crowdsourced weather predictions for specific locations, where information is drawn from the barometric pressure and temperature readings of users' smartphones (since launched as the WeatherSignal app), and how they were exploring the possibilities of correlating indoor and outdoor sound discrepancies with Wi-Fi signal strength to improve weather readings and determine when a phone user was moving indoors (Khanifar, pers. comm., 2013).

Location has also featured heavily in a number of recent corporate applications of machine learning that employ "intricate data practices— normalization, regularization, cross-validation, feature engineering, feature selection, optimization—[in order to] embroider datasets into shapes they can recognize" (Mackenzie 2017, 5). For instance, Google applies machine-learning techniques to newly recorded Street View footage. This footage is then analyzed in order to extract from this footage "the street names and numbers, and properly create and locate the new addresses, automatically, on Google Maps" (quoted in Buczkowski 2017a).

And, finally, there is also strong, emerging interest in location among the blockchain developer community. One Ethereum blockchain platform start-up, FOAM, has developed what they call "proof of location," "a cryptographic method for proving that a user has actually been at a certain location," which, the FOAM developers argue, addresses a known problem affecting a range of location-related services (Hertig 2018).

In writing this book, I have sought to develop as full and rich a picture as possible of the present state of play of locative media, and the complicated cultural economies associated with their ownership and operation and their personal and sociocultural use. I have also sought to provide a sound basis for assessing the extent to which the rise of location-awareness and location-based services have realized significant shifts in the way we presently understand the development, consumption, business operations, and policy and regulatory implications of new media technologies. All of this work is complicated by the fact that "complex objects such as media systems" only ever settle "temporarily into what passes for a stable state" before reforming and resettling, and so on, in a process that is ongoing (Fuller 2005, 1). To put the challenge that is before us in the language of locative media and mapping, in analyzing these services and their uses, the critical pins we are trying to drop are situated in a fluid and ever-shifting technological landscape. Thus, while location continues to hold a central place in the new technological directions and developments described in this Conclusion, these also throw up a whole new set of critical questions, and suggest the need for reinvigorated theoretical approaches (Wilken 2018b) and fresh critical tools. If we are to make critical sense of the continuing importance, as well as the possible future implications and impacts, of location, location-awareness, and location-based services, exploring these questions and developing these tools are the tasks now at hand.

REFERENCES

4sweep. 2018. "4sweep: A Power Tool for Foursquare Superusers." *foursweep.com*.
    https://foursweep.com

140Talks. 2011. "#140 conf NYC 2011: 'QA with Dennis Crowley.'" *YouTube*, December
    20. https://youtu.be/K-CdwtX0ci0

Aas, Katja Franko. 2013. "'Getting Ahead of the Game': Border Technologies and the
    Changing Space of Governance." In *Global Surveillance and Policing: Borders,
    Security, Identity*, edited by Elia Zureik and Mark B. Salter, 194–214.
    London: Routledge.

Abdesslem, Fehmi Ben, Iain Parris, and Tristan Henderson. 2010. "Mobile Experience
    Sampling: Reaching the Parts of Facebook Other Methods Cannot Reach."
    Proceedings of the Privacy and Usability Methods Pow-Wow (PUMP), Dundee,
    UK (British Computer Society, September, 2010), 1–8. http://scone.cs.st-
    andrews.ac.uk/pump2010/papers/benabdesslem.pdf

Abidin, Crystal. 2016. "Visibility Labour: Engaging with Influencers' Fashion Brands
    and #OOTD Advertorial Campaigns on Instagram." *Media International
    Australia* 161: 86–100.

Abidin, Crystal. 2017. "Influencer Extravaganza: A Decade of Commercial 'Lifestyle'
    Microcelebrities in Singapore." In *Routledge Companion to Digital Ethnography*,
    edited by Larissa Hjorth, Heather Horst, Genevieve Bell and Anne Galloway,
    158–168. London: Routledge.

"About Foursquare." 2013. *Foursquare*, September. https://foursquare.com/about

"About Us." 2018. *Foursquare*. https://foursquare.com/about

Aceti, Lanfranco, Hana Iverson, and Mimi Sheller, eds. 2016. "L.A. Re.Play: Mobile
    Network Culture in Placemaking," special themed issue of *Leonardo Electronic
    Almanac* 21 (1). http://www.leoalmanac.org/l-a-re-play-volume-21-no-1/

Adams, Paul C. 2009. *Geographies of Media and Communication*. Chichester, West
    Sussex: Wiley-Blackwell.

Agamben, Giorgio. 2001. "Security and Terror." Translated by Carolin Emcke. *Theory
    and Event* 5 (4). https://doi.org/10.1353/tae.2001.0030

Agamben, Giorgio. 2005. *State of Exception*. Translated by Kevin Attell.
    Chicago: University of Chicago Press.

Ahmed, Nafeez. 2015a. "How the CIA Made Google: Inside the Secret
    Network Behind Mass Surveillance, Endless War, and Skynet—Part 1."
    Medium, January 23. https://medium.com/insurge-intelligence/
    how-the-cia-made-google-e836451a959e

Ahmed, Nafeez. 2015b. "Why Google Made the NSA: Inside the Secret Network Behind Mass Surveillance, Endless War, and Skynet—Part 2." *Medium*, January 22. https://medium.com/insurge-intelligence/why-google-made-the-nsa-2a80584c9c1

Aked, Alan. 1990. "AUSSAT as a Telecommunications Satellite." *Media Information Australia* 58 (1): 55–59.

Akil, Omari. 2016. "Warning: Pokémon Go Is a Death Sentence if You Are a Black Man." *Medium*, July 7. https://medium.com/mobile-lifestyle/warning-pokemon-go-is-a-death-sentence-if-you-are-a-black-man-acacb4bdae7f

Albarran, Alan B., ed. 2013. *The Social Media Industries*. New York: Routledge.

Albury, Kath, Jean Burgess, Ben Light, Kane Race, and Rowan Wilken. 2017. "Data Cultures of Mobile Dating and Hook-Up Apps: Emerging Issues for Critical Social Science Research." *Big Data & Society*. https://doi.org/10.1177/2053951717720950

Albury, Kath, and Paul Byron. 2016. "Safe on My Phone? Same-Sex Attracted Young People's Negotiations of Intimacy, Visibility, and Risk on Digital Hook-up Apps." *Social Media + Society* (October–December). https://doi.org/10.1177/2056305116672887

Allan, Alasdair. 2011. "Got an iPhone or 3G iPad? Apple Is Recording Your Moves." *O'Reilly Radar*, April 20. http://radar.oreilly.com/2011/04/apple-location-tracking.html#whats-happening

Alphabet Inc. 2014. "Google and Skybox Imaging Sign Acquisition Agreement." *Alphabet Investor Relations*, June 10. https://abc.xyz/investor/news/releases/2014/0609.html

Altheide, David L. 1994. "An Ecology of Communication: Toward a Mapping of the Effective Environment." *The Sociological Quarterly* 35 (4): 665–683.

Altheide, David L. 1995. *An Ecology of Communication: Cultural Formats of Control*. New York: Aldine de Gruyter.

Altman, Irwin. 1975. *The Environment and Social Behavior: Privacy, Personal Space, Territory, Crowding*. Monterey, CA: Brooks/Cole.

Altman, Irwin. 1977. "Privacy Regulation: Culturally Universal or Culturally Specific?" *Journal of Social Issues* 33 (3): 66–84.

Altman, Irwin. 1990. "Toward a Transactional Perspective: A Personal Journey." In *Environment and Behavior Studies: Emergence of Intellectual Traditions*, edited by Irwin Altman and Kathleen Christensen, 225–255. New York: Plenum Press.

Alvarez León, Luis F. 2016. "Property Regimes and the Commodification of Geographic Information: An Examination of Google Street View." *Big Data & Society* (July–December). https://doi.org/10.1177/2053951716637885/

Alvarez León, Luis F. 2019. How Cars Became Mobile Spatial Media: A Geographical Political Economy of On-Board Navigation." *Mobile Media & Communication*. https://doi.org/10.1177/2050157919826356

Ambrosia, Vince, Mike Hutt and Kamlesh Lulla. 2011. "Editorial." *Geocarto International* 26 (2): 69–70.

Amin, Ash, and Nigel Thrift. 2004. "Introduction." In *The Blackwell Cultural Economy Reader*, edited by Ash Amin and Nigel Thrift, x–xxx. Malden, MA: Blackwell.

Ananny, Mike, and Kate Crawford. 2018. "Seeing Without Knowing: Limitations of the Transparency Ideal and Its Application to Algorithmic Accountability," *New Media & Society* 20 (3): 973–989.

Andrejevic, Mark. 2007. "Surveillance in the Digital Enclosure." *The Communication Review* 10: 295–317.

Andrejevic, Mark. 2010. "Social Network Exploitation." In *A Networked Self: Identity, Community, and Culture on Social Network Sites*, edited by Zizi Papacharissi, 82–101. New York: Routledge.

Andrejevic, Mark. 2015. "Becoming Drones: Smartphone Probes and Distributed Sensing." In *Locative Media*, edited by Rowan Wilken and Gerard Goggin, 193–207. New York: Routledge.

Andrejevic, Mark, and Mark Burdon. 2015. "Defining the Sensor Society." *Television & New Media*. 16 (1): 19–36.

Angwin, Julia, and Jeff Larson. 2014. "The NSA Revelations All in One Chart." *ProPublica*, June 30. https://projects.propublica.org/nsa-grid/

Angwin, Julia, Charlie Savage, Jeff Larson, Henrik Moltke, Laura Poitras, and James Risen. 2015. "AT&T Helped US Spy on Internet on a Vast Scale." *New York Times*, August 15. https://www.nytimes.com/2015/08/16/us/politics/att-helped-nsa-spy-on-an-array-of-internet-traffic.html?mcubz=1

Anheier, Helmut, and Yudhishthir Raj Isar. 2008. "Introducing *The Cultures of Globalization Series* and *The Cultural Economy*." In *The Cultural Economy*, edited by Helmut Anheier and Yudhishthir Raj Isar, 1–12. London: SAGE.

"Announcing Button & Foursquare." 2015. *Button.com*, June 1. http://building.usebutton.com/announcement/2015/06/01/button-foursquare-uber/

Apple. 2018. "Using the Significant-Change Location Service." *Apple Developer*. https://developer.apple.com/documentation/corelocation/getting_the_user_s_location/using_the_significant_change_location_service

"Apple Kills Drone Strike News App for 'Objectionable Content.'" 2015. *RT.com*, September 29. http://www.rt.com/usa/316955-apple-ends-drone-app/

"Apple Reports Third Quarter Results." 2012. *Apple Newsroom*, July 24. https://www.apple.com/pr/library/2012/07/24Apple-Reports-Third-Quarter-Results.html

Arminen, Ilkka. 2006. "Social Functions of Location in Mobile Telephony." *Personal and Ubiquitous Computing* 10: 319–323.

Arthur, Charles. 2009. "Why Did Apple Buy the Mapping Company Placebase?" *The Guardian* October 2. http://www.theguardian.com/technology/blog/2009/oct/01/apple-maps-placebase-google-question

Arvidsson, Adam. 2006. *Brands: Meaning and Value in Media Culture*. London: Routledge.

Aslinger, Ben. 2012. "WorldSpace Satellite Radio and the South African Footprint." In *Down to Earth: Satellite Technologies, Industries, and Cultures*, edited by Lisa Parks and James Schwoch, 194–203. New Brunswick, NJ: Rutgers University Press.

Auto Channel. 2001. "Telcontar Announces Close of $23.6 Million Funding Round; Strategic Investors in Location-Based Software & Services Company Include Ford Motor Company." *The Auto Channel*, March 20. http://www.theautochannel.com/news/2001/03/20/017039.html

Bachelard, Gaston. 1994. *The Poetics of Space*. Translated by Maria Jolas. Boston: Beacon Press.

Balibar, Étienne. 1998. "The Borders of Europe." Translated by James Swenson. In *Cosmopolitics: Thinking and Feeling Beyond the Nation*, edited by Pheng Cheah and Bruce Robbins, 216–229. Minneapolis: University of Minnesota Press.

Ball, James. 2013. "US and UK Struck Secret Deal to Allow NSA to 'Unmask' Britons' Personal Data." *The Guardian*, November 21. https://www.theguardian.com/world/2013/nov/20/us-uk-secret-deal-surveillance-personal-data

Ball, James. 2014a. "Angry Birds and "Leaky" Phone Apps Targeted
by NSA and GCHQ for User Data." *The Guardian*, January
28. https://www.theguardian.com/world/2014/jan/27/
nsa-gchq-smartphone-app-angry-birds-personal-data

Ball, James. 2014b. "NSA Collects Millions of Text Messages
Daily in 'Untargeted' Global Sweep." *The Guardian*, January
17. https://www.theguardian.com/world/2014/jan/16/
nsa-collects-millions-text-messages-daily-untargeted-global-sweep

Ball, James. 2014c. "NSA Stores Metadata of Millions of Web Users
for Up to a Year, Secret Files Show." *The Guardian*, October
1. https://www.theguardian.com/world/2013/sep/30/
nsa-americans-metadata-year-documents

Ball, James, Luke Harding, and Juliette Garside. 2013. "BT and Vodafone
among Telecoms Companies Passing Details to GCHQ." *The Guardian*,
August 3. https://www.theguardian.com/business/2013/aug/02/
telecoms-bt-vodafone-cables-gchq

Bambozzi, Lucas. 2009. "Risky Approximations between Site-Specific and Locative
Arts." *Wi: Journal of Mobile Media* 10. http://wi.hexagram.ca/?p=56

Bamford, James. 1982. *The Puzzle Palace: A Report on America's Most Secret Agency*.
Boston: Houghton Mifflin.

Bamford, James. 2001. *Body of Secrets: Anatomy of the Ultra-secret Security Agency*.
New York: Doubleday.

Bamford, James. 2005. "The Agency That Could Be Big Brother." *New York Times*,
December 25. http://www.nytimes.com/2005/12/25/weekinreview/the-
agency-that-could-be-big-brother.html

Bamford, James. 2008. *The Shadow Factory: The Ultra-secret NSA from 9/11 to the
Eavesdropping on America*. New York: Doubleday.

Bamford, James. 2012. "The NSA Is Building the Country's Biggest Spy Center
(Watch What You Say)." *Wired*, March 15. https://www.wired.com/2012/03/ff_
nsadatacenter/

Bann, Stephen. 1995. "Nature Over Again after Poussin: Some Discovered
Landscapes." In *Wood Notes Wild: Essays on the Poetry and Art of Ian Hamilton
Finlay*, edited by Alec Finlay, 98–115. Edinburgh: Polygon.

Barbour, Kim. 2015. "Registers of Performance: Negotiating the Professional,
Personal and Intimate in Online Persona Creation." In *Media, Margins and
Popular Culture*, edited by Einar Thorsen, Heather Savigny, Jenny Alexander
and Daniel Jackson, 57–69. London: Palgrave Macmillan.

Barouch, Jonathan. 2013. "Foursquare's API Is a Pillar of the Mobile App
Ecosystem." *TechCrunch*, March 29. http://techcrunch.com/2013/03/29/
the-internet-needs-foursquare-to-succeed/

Barreneche, Carlos. 2012a. "Governing the Geocoded World: Environmentality and
the Politics of Location Platforms." *Convergence: The International Journal of
Research into New Media Technologies* 18 (3): 331–351.

Barreneche, Carlos. 2012b. "The Order of Places: Code, Ontology and Visibility in
Locative Media." *Computational Culture: A Journal of Software Studies* 2. http://
computationalculture.net

Barreneche, Carlos. 2015. "The Cluster Diagram: A Topological Analysis of Locative
Networking." In *Locative Media*, edited by Rowan Wilken and Gerard Goggin,
107–117. New York: Routledge.

Barreneche, Carlos, and Rowan Wilken. 2015. "Platform Specificity and the
    Politics of Location Data Extraction." *European Journal of Cultural Studies* 18
    (4–5): 497–513.

Baudrillard, Jean. 1995. *The Gulf War Did Not Take Place*. Translated by Paul Patton.
    Sydney: Power Publications.

Baym, Nancy. 2010. *Personal Connections in the Digital Age*. Cambridge: Polity.

Baym, Nancy. 2011. "Introducing Facebook Nation." *Social Media Collective
    Research Blog*, September 27. http://socialmediacollective.org/2011/09/27/
    introducing-facebook-nation/

Bazarova, Natalya N. 2012. "Public Intimacy: Disclosure Interpretation and Social
    Judgments on Facebook." *Journal of Communication* 62 (5): 815–832.

Beck, Ulrich. 1994. "The Reinvention of Politics: Towards a Theory of Reflexive
    Modernization." In *Reflexive Modernization*, edited by Ulrich Beck, Anthony
    Giddens, and Scott Lash, 1–55. Cambridge: Polity.

Beck, Ulrich. 1999. *World Risk Society*. Malden, MA: Polity Press.

Beck, Ulrich, and Elisabeth Beck-Gernsheim. 2002. *Individualization: Institutionalized
    Individualism and its Social and Political Consequences*. London: SAGE.

Beer, David. 2016. *Metric Power*. Houndmills, Basingstoke: Palgrave Macmillan.

Begley, Josh. 2015. "The Drone Papers, Article No. 2 of 8: A Visual Glossary—
    Decoding the Language of Covert Warfare." *The Intercept*, October 15. https://
    theintercept.com/drone-papers/a-visual-glossary/

Behrendt, Frauke. 2012. "The Sound of Locative Media." *Convergence: The International
    Journal of Research into New Media Technologies* 18 (3): 283–295.

Bellanova, Rocco. 2011. "Waiting for the Barbarians or Shaping New Societies?"
    Information Polity 16: 391–395.

Bellanova, Rocco. 2017. "Digital, Politics, and Algorithms: Governing Digital Data
    through the Lens of Data Protection." *European Journal of Social Theory* 20
    (3): 329–347.

Benford, Steve, Chris Greenhalgh, Gabriella Giannachi, Brendan Walker, Joe
    Marshall, and Tom Rodden. 2012. "Uncomfortable Interactions." CHI'12,
    Austin, Texas, May 5–10.

Benson-Allott, Caetlin. 2013. *Killer Tapes and Shattered Screens: Video Spectatorship
    from VHS to File Sharing*. Berkeley: University of California Press.

Benthall, Sebastian, Seda Gürses, and Helen Nissenbaum. 2017. "Contextual Integrity
    through the Lens of Computer Science." *Foundations and Trends in Privacy and
    Security* 2 (1): 1–69.

Bercovici, Jeff. 2013. "Apple and Topsy: Unpeeling the $200 Million Mystery."
    Forbes, December 4. http://www.forbes.com/sites/jeffbercovici/2013/12/04/
    apple-and-topsy-unpeeling-the-200-million-mystery/

Bercovici, Jeff. 2014. "LinkedIn for Love, Tinder for Business and Other Off-label
    Technology Uses." *Forbes*, May 6. https://www.forbes.com/sites/jeffbercovici/
    2014/05/06/linkedin-for-love-tinder-for-business-and-other-off-label-
    technology-uses/#3895ffa62e41

Bergen, Mark, and Dawn Chmielewski. 2015. "Apple Acquires Mapsense, a Mapping
    Visualization Startup." *re/code*, September 16. http://recode.net/2015/09/16/
    apple-acquires-mapsense-a-mapping-visualization-startup/

Berger, John. 2001. *The Shape of a Pocket*. New York: Vintage.

Berger, John. 2016. *Landscapes: John Berger on Art*. London: Verso.

Berry, Marsha. 2017. *Creating with Mobile Media*. New York: Palgrave Macmillan.

Berry, Marsha, and Omega Goodwin. 2013. "Poetry 4 U: Pinning Poems under/over/
through the Streets." *New Media & Society* 15 (6): 909–929.

Berry, Marsha, and Max Schleser, eds. 2014. *Mobile Media Making in an Age of
Smartphones*. New York: Palgrave Macmillan.

Bertel, Troels Fibæk. 2016. "'Why Would You Want to Know?': The Reluctant
Use of Location Sharing via Check-ins on Facebook among Danish Youth."
Convergence: The International Journal of Research into New Media Technologies
22 (2): 162–176.

Best, Jacqueline, and Matthew Paterson. 2010. "Introduction: Understanding
Cultural Political Economy." In *Cultural Political Economy*, edited by Jacqueline
Best and Matthew Paterson, 1–25. Hoboken, NJ: Routledge.

Bhasker, Michael. 2016. *Curation: The Power of Selection in a World of Excess*.
London: Piatkus.

Biggs, Tim. 2017. "Federal Court Rejects Application for Telstra to Supply 'Personal'
Metadata." *smh.com.au*, January 20. https://www.smh.com.au/technology/
federal-court-rejects-application-for-telstra-to-supply-personal-metadata-
20170120-gtvc85.html

Bilton, Nick 2010. "Facebook Will Allow Users to Share Location." *New York Times*,
March 9. http://bits.blogs.nytimes.com/2010/03/09/facebook-will-allow-
users-to-share-location/?_r=0

Blanchette, Jean-François, and Deborah G. Johnson. 2002. "Data Retention and the
Panoptic Society: The Social Benefits of Forgetfulness." *The Information Society*
18: 33–45.

Blank, Steve. 2010. "What's a Start-up? First Principles." *Steve Blank*, January 25.
http://steveblank.com/2010/01/25/whats-a-startup-first-principles/

Blast Theory. 2008. "You Get Me." *Blast Theory*. http://www.blasttheory.co.uk/
projects/you-get-me/

Blast Theory. 2016. "Our History & Approach." *Blast Theory*. http://
www.blasttheory.co.uk/our-history-approach

Bleeker, Julian, and Jeff Knowlton. 2006. "Locative Media: A Brief Bibliography and
Taxonomy of GPS-Enabled Locative Media." *Leonardo Electronic Almanac* 14 (3).
http://leoalmanac.org/journal/vol_14/lea_v14_n03-04/jbleecker.html

Block, Ryan. 2005. "Google Buys Dodgeball." *Engadget*, May 12. https://
www.engadget.com/2005/05/12/google-buys-dodgeball/

Blumberg, Andrew J., and Peter Eckersley. 2009. "On Locational Privacy and How
to Avoid Losing It Forever." *Electronic Frontier Foundation*, August. http://
www.eff.org/wp/locational-privacy

Blumenthal, Mike. 2009. "Tectonic Shifts Altering the Terrain at Google
Maps." *Search Engine Land*, October 14. https://searchengineland.com/
tectonic-shifts-altering-the-terrain-at-google-maps-27783

Bollmer, Grant. 2016. *Inhuman Networks: Social Media and the Archaeology of
Connection*. New York: Bloomsbury.

Bonnett, Alistair. 2014. *Off the Map: Lost Spaces, Invisible Cities, Forgotten Islands,
Feral Places, and What They Tell Us about the World*. London: Aurum Press.

Bowker, Geoffrey C., Karen Baker, Florence Millerand, and David Ribes. 2010.
"Toward Information Infrastructure Studies: Ways of Knowing in a Networked
Environment." In *International Handbook of Internet Research*, edited by Jeremy
Hunsinger, Lisbeth Klastrup and Matthew Allen, 97–117. Dordrecht: Springer.

Bowker, Geoffrey C., and Susan Leigh Star. 2000. *Sorting Things Out: Classification and
Its Consequences*. Cambridge, MA: MIT Press.

boyd, danah, and Alice Marwick. 2011. "Social Privacy in Networked Publics: Teens' Attitudes, Practices, and Strategies." Paper presented at Privacy Law Scholars Conference, June 2, Berkeley, CA.

Boyle, Casey. 2015. "The Rhetorical Question Concerning Glitch." *Computers and Composition* 35: 12–29.

Braman, Sandra. 2009. *Change of State: Information, Policy, and Power*. Cambridge, MA: MIT Press.

Brassey, Dom (@domlet). 2018. "Every Data Set Has a Spatial Component." Twitter, May 31, 2018, 7:36 a.m. https://twitter.com/domlet/status/1001940322946093058

Braue, David. 2013. "Apple Maps' Worldview Is Now Better Than Google Maps." *ZDNet*, November 30. http://www.zdnet.com/article/apple-maps-worldview-is-now-better-than-google-maps/

Braun, Joshua A. 2015. *This Program Is Brought to You by . . . Distributing Television News Online*. New Haven, CT: Yale University Press.

Bray, Hiawatha. 2014. "Court Tosses Skyhook's Suit against Google." *Boston Globe*, November 6. http://www.bostonglobe.com/business/2014/11/06/court-tosses-local-suit-against-google/fJK8HLbiv8pJoGM1TfIxiP/story.html

Brock, André. 2009. "'Who Do You Think You Are?': Race, Representation, and Cultural Rhetorics in Online Spaces." *POROI: An Interdisciplinary Journal of Rhetorical Analysis & Invention* 6 (1): 15–35.

Brock, André. 2012. "From the Blackhand Side: Twitter as a Cultural Conversation." *Journal of Broadcasting & Electronic Media* 56 (4): 529–549.

Brodmerkel, Sven, and Nicholas Carah. 2016. *Brand Machines, Sensory Media and Calculative Culture*. London: Palgrave Macmillan.

Brownlee, John. 2012. "This Creepy App Isn't Just Stalking Women without Their Knowledge." *Cult of Mac*. http://www.cultofmac.com/157641/this-creepy-app-isnt-just-stalking-women-without-their-knowledge-its-a-wake-up-call-about-facebook-privacy/

Bucher, Taina. 2011. "Network as Material: An Interview with Julian Oliver." *Furtherfield*, June 16. http://www.furtherfield.org/features/network-material-interview-julian-oliver

Bucher, Taina. 2012a. "A Technicity of Attention: How Software 'Makes Sense.'" *Culture Machine* 13: 1–23. http://www.culturemachine.net/index.php/cm/article/view/470/489

Bucher, Taina. 2012b. "Want to Be on the Top? Algorithmic Power and the Threat of Invisibility on Facebook." *New Media & Society* 14 (7): 1164–1180.

Bucher, Taina. 2013a. "Objects of Intense Feeling: The Case of the Twitter API." *Computational Culture: A Journal of Software Studies* (3). http://computationalculture.net/article/objects-of-intense-feeling-the-case-of-the-twitter-api

Bucher, Taina. 2013b. "The Friendship Assemblage: Investigating Programmed Sociality on Facebook." *Television & New Media* 14 (6): 495–509.

Bucher, Taina. 2017. "The Algorithmic Imaginary: Exploring the Ordinary Affects of Facebook Algorithms." *Information, Communication & Society* 17 (1): 30–44.

Bucher, Taina, and Anne Helmond. 2018. "The Affordances of Social Media Platforms." In *The SAGE Handbook of Social Media*, edited by Jean Burgess, Thomas Poell and Alice Marwick, 233–253. London: SAGE.

Buczkowski, Aleks. 2015. "Everything There Is to Know about Apple Mapping Vans." *Geoawesomeness*, August 11. http://geoawesomeness.com/everything-there-is-to-know-about-apple-mapping-vans/

Buczkowski, Aleks. 2016a. "Alibaba Invests in PlaceIQ—Location-Based Targeting Start-up." *Geoawesomeness*, October 18. https://geoawesomeness.com/alibaba-invests-placeiq-biggest-location-based-targeting-start/

Buczkowski, Aleks. 2016b. "List of the Top 100 Geospatial Start-ups and Companies in the World." *Geoawesomeness*, February 11. http://geoawesomeness.com/list-top-100-geospatial-start-ups-companies-world/#comment-2508535745

Buczkowski, Aleks. 2017a. "Google Maps Goes on Deep Learning Diet to Update Addresses." *Geoawesomeness*, May 5. http://geoawesomeness.com/google-maps-goes-deep-learning-diet-update-addresses/

Buczkowski, Aleks. 2017b. "Sundar Pichai Tells How Google Plans to Make Money Off Maps." *Geoawesomeness*, April 28. http://geoawesomeness.com/sundar-pichai-tells-google-plans-make-money-off-maps/

Buczkowski, Aleks. 2018. "After Launching Its First Fully Owned Maps This Week, Apple Is Now Officially 'a Mapping Company.'" *Geoawesomeness*, July 4. http://geoawesomeness.com/apple-is-now-officially-a-mapping-company/

Bull, Michael. 2007. *Sound Moves: iPod Culture and Urban Experience*. London: Routledge.

Bunz, Mercedes, and Graham Meikle. 2018. *The Internet of Things*. Cambridge: Polity.

Burchell, Kenzie. 2017. "Finding Time for Goffman: When Absence Is More Telling than Presence." In *Conditions of Mediation: Phenomenological Perspectives on Media*, edited by Tim Markham and Scott Rodgers, 185–195. New York: Berg.

Burdon, Mark, and Alissa McKillop. 2013. "The Google Street View Wi-Fi Scandal and Its Repercussions for Privacy Regulation." *Monash University Law Review* 39 (3): 702–738.

Bureau. 2017. "Drone Warfare." *The Bureau of Investigative Journalism*. https://www.thebureauinvestigates.com/projects/drone-war

Burns, Kelli S. 2013. "Self-presentation and Privacy Considerations of Foursquare Users." In *Mobile Media Practices, Presence and Politics: The Challenge of Being Seamlessly Mobile*, edited by Kathleen M. Cumiskey and Larissa Hjorth, 150–165. New York: Routledge.

Burns, Patrick L. 2012. "SXSW 2012: The Year of the Ambient Social Location App." *arc3|communications*, March 27. http://arc3communications.com/sxsw-2012-the-year-of-the-ambient-social-location-app/

Burrows, Peter, and Sarah Frier. 2013. "Apple Said to Buy HopStop, Pushing Deeper into Maps." *Bloomberg*, July 21. http://www.bloomberg.com/news/articles/2013-07-19/apple-said-to-buy-hopstop-pushing-deeper-into-maps

Bushee, Brian J., John E. Core, Wayne Guay, and Sophia J. W. Hamm. 2010. "The Role of the Business Press as an Information Intermediary." *Journal of Accounting Research* 48 (1): 1–19.

Butler, Judith. 1998. "Performative Acts and Gender Constitution: An Essay in Phenomenology and Feminist Theory." *Theatre Journal* 40 (4): 519–531.

Butler, Judith. 1990. *Gender Trouble: Feminism and the Subversion of Identity*. New York: Routledge.

Butler, Judith. 2004. *Precarious Life: The Powers of Mourning and Violence*. London: Verso.

Bygrave, Lee A. 2017. "Hardwiring Privacy." In *The Oxford Handbook of Law, Regulation, and Technology*, edited by Roger Brownsword, Eloise Scotford and Karen Yeung, 754–775. Oxford: Oxford University Press.

Calore, Michael. 2013. "How Foursquare Is Forcing Social Networks to Check In or Check Out." *Wired*, March 12. http://www.wired.com/underwire/2013/03/location-apps-social-media/

Campbell, Duncan. 2015. "GCHQ and Me: My Life Unmasking British Eavesdroppers." *The Intercept*, August 3. https://theintercept.com/2015/08/03/life-unmasking-british-eavesdroppers/

Captain, Sean. 2017. "Qualcomm's New Camera Will Give Smartphones 3D Vision." *Fast Company*, August 15. https://www.fastcompany.com/40451913/qualcomms-new-camera-will-give-smartphones-3d-vision

Carah, Nicholas. 2014. "Curators of Databases: Circulating Images, Managing Attention and Making Value on Social Media." *Media International Australia* 150: 137–142.

Carlson, Anna. 2018. "[BigDataSur] blog 1/3: Imagining 'Good' Data: Northern Utopias, Southern Lessons." *Datactive: The Politics of Data According to Civil Society*, May 25. https://data-activism.net/2018/05/bigdatasur-blog-13-imagining-good-data-northern-utopias-southern-lessons/

Carr, Austin. 2014. "Instagram Testing Facebook Places Integration to Replace Foursquare." *Fast Company*, March 25. https://www.fastcompany.com/3028166/instagram-testing-facebook-places-integration-to-replace-foursquare

Castells, Manuel, Mireia Fernández-Ardèvol, Jack Linchuan Qiu, and Araba Sey. 2007. *Mobile Communication and Society: A Global Perspective*. Cambridge, MA: MIT Press.

Ceruzzi, Paul E. 2018. *GPS*. Cambridge, MA: MIT Press.

Chacko, Marissa. 2016. "Introducing Marsbot." *Foursquare Blog*. http://blog.foursquare.com/post/144872708248/introducing-marsbot

Chakravarty, Shilpi. 2018. "How GDPR Will Impact Location Data." *Geospatial World*, May 23. https://www.geospatialworld.net/article/how-gdpr-impacts-location-data/

Chamberlain, Kerry, and Antonia C. Lyons. 2016. "Using Material Objects and Artifacts in Research." In *Routledge Handbook of Qualitative Research in Sport and Exercise*, edited by Brett Smith and Andrew C. Sparkes, 164–177. London: SAGE.

Chang, Chen-Wei, and Gina Masullo Chen. 2014. "College Students' Disclosure of Location-Related Information on Facebook." *Computers in Human Behavior* 35: 33–38.

Chang, Emily, and Douglas MacMillan. 2011. "Foursquare Says Merchant Services Will Provide Bulk of Revenue." *Bloomberg*, August 2. http://www.bloomberg.com/news/2011-08-02/foursquare-ceo-says-merchant-services-will-provide-bulk-of-startup-s-sales.html

Chaturvedi, Aditya. 2018. "How Facebook Is Attempting to Circumvent GDPR." *Geospatial World*, May 24. https://www.geospatialworld.net/article/facebook-is-not-yielding/

Chen, Brian X. 2017. "The Smartphone's Future: It's All about the Camera." *New York Times*, August 30. https://www.nytimes.com/2017/08/30/technology/personaltech/future-smartphone-camera-augmented-reality.html

Chow, Raymond. 2013. "Why-Spy? An Analysis of Privacy and Geolocation in the Wake of the 2010 Google 'Why-Spy' Controversy." *Rutgers Computer and Technology Law Journal* 39: 56–94.

Clark, Thomas A. 1995. "Pastorals." In *Wood Notes Wild: Essays on the Poetry and Art of Ian Hamilton Finlay*, edited by Alec Finlay, 152–155. Edinburgh: Polygon.

Clayton, Mark. 2013. "NSA Surveillance: Revelations Damaged US
Security, Obama Says." *The Christian Science Monitor*, December 20.
https://www.csmonitor.com/World/Security-Watch/2013/1220/
NSA-surveillance-Revelations-damaged-US-security-Obama-says

Cleland, Scott. 2010. "Google Wi-Spy Was an Intentional Plan to Beat Skyhook
Wireless." *Precursor Blog*, November 9. http://www.precursorblog.com/
?q=content/google-wi-spy-was-intentional-plan-beat-skyhook-wireless

Clover, Juli. 2014. "Apple Hires Developers Behind Defunct 'Pin Drop' Mapping
App." *MacRumors*, November 10. http://www.macrumors.com/2014/11/10/
apple-pin-drop-maps-hires/

Clucas, Stephen. 2000. "Cultural Phenomenology and the Everyday." *Critical
Quarterly* 42 (1): 8–34.

CNIL. 2011. "Délibération n°2011-035 du 17 mars 2011 de la formation restreinte
prononçant une sanction pécuniaire à l'encontre de la société X." Commission
Nationale de l'Informatique et des Libertés (CNIL), March 17. https://
www.legifrance.gouv.fr/affichCnil.do?&id=CNILTEXT000023733987

Coast, Steve. 2011. "How OpenStreetMap Is Changing the World." In *Web and Wireless
Geographical Information Systems, 10th International Symposium, W2GIS 2011,
Kyoto, Japan, March 3–4, 2011, Proceedings*, edited by Katshushi Tanaka, Peter
Fröhlich and Kyoung-Sook Kim. Heidelberg: Springer.

Coast, Steve. 2015. *The Book of OSM*. Lexington, KY: CreateSpace Independent
Publishing Platform.

Cohen, Julie E. 2012. *Configuring the Networked Self: Law, Code, and the Play of
Everyday Practice*. New Haven, CT: Yale University Press.

Cole, Teju. 2015. "Officer Involved." *The Intercept*. https://theintercept.co/
officer-involved/

Coleman, David J., and John D. McLaughlin. 1998. "Defining Global Geospatial Data
Infrastructure (GGDI): Components, Stakeholders and Interfaces." *Geomatica*
52 (2): 129–143.

Collis, Christy. 2012. "The Geostationary Orbit: A Critical Legal Geography of Space's
Most Valuable Real Estate." In *Down to Earth: Satellite Technologies, Industries,
and Cultures*, edited by Lisa Parks and James Schwoch, 61–81. New Brunswick,
NJ: Rutgers University Press.

Constine, Josh. 2010. "Facebook Updates Mobile Platform with Location APIs, Single
Sign-on, and Deals." *Adweek*, November 3. https://www.adweek.com/digital/
launches-mobile-platform-location/

Constine, Josh. 2011a. "Facebook Asks Users to Clean Up Its Location
Database with Places Editor and Favorite Places." *Inside
Facebook*, July 1. http://www.insidefacebook.com/2011/07/01/
favorite-places-editor-location-database/

Constine, Josh. 2011b. "Facebook Overhauls Privacy, Brings Control In-line
with Content." *Adweek*, August 23. https://www.adweek.com/digital/
in-line-privacy/

Constine, Josh. 2012a. "800K #Sandy-grams Showed Systrom Instagram Is 'Going
to Need to Be a Big Data Company.'" *TechCrunch*, November 6. http://
techcrunch.com/2012/11/05/instagram-big-data/

Constine, Josh. 2012b. "How Big Is Facebook's Data? 2.5 Billion Pieces of Content
and 500+ Terabytes Ingested Every Day." *TechCrunch*, August 22. https://
techcrunch.com/2012/08/22/how-big-is-facebooks-data-2-5-billion-pieces-of-
content-and-500-terabytes-ingested-every-day/

Constine, Josh. 2012c. "Josh Williams on Why Facebook Acquired His Startup Gowalla to Put a 'Nearby' Lens on 250M Geo-posts per Month." *TechCrunch*, December 17. http://techcrunch.com/2012/12/17/facebook-nearby-gowalla/

Constine, Josh. 2012d. "Facebook Becomes Location Backbone That Lets Apps Import Checkins from Each Other." *TechCrunch*, March 7. http://techcrunch.com/2012/03/07/facebook-becomes-location-backbone-that-lets-apps-import-checkins-from-each-other/

Constine, Josh. 2012e. "Facebook SoLoMo-fies the Platform, Lets App Tag Friends and Places." *TechCrunch*, March 7. http://techcrunch.com/2012/03/07/facebook-solomo-fies-the-platform-lets-apps-tag-friends-and-places/

Constine, Josh. 2012f. "Hands On with Facebook Nearby, a New Local Biz Discovery Feature That Challenges Yelp and Foursquare." *TechCrunch*, December 18. https://techcrunch.com/2012/12/17/facebook-nearby/

Constine, Josh. 2014. "Facebook Launches 'Nearby Friends' with Opt-in Real-time Location Sharing to Help You Meet Up." *TechCrunch*, April 18. https://techcrunch.com/2014/04/17/facebook-nearby-friends/

Constine, Josh. 2015. "Facebook Messenger Ditches Constant Mapping to Lay Groundwork for More Location Features." *TechCrunch*, June 5. https://techcrunch.com/2015/06/04/foursquessenger/

Constine, Josh. 2016. "Facebook Kills Off Exact Location Sharing in Nearby Friends, Adds 'Wave.'" *TechCrunch*, December 23. https://techcrunch.com/2016/12/22/facebook-wave/

Constine, Josh, and Kim-Mai Cutler. 2012. "Facebook Buys Instagram for $1 Billion, Turns Budding Rival into Its Standalone Photo App." *TechCrunch*, April 9. http://techcrunch.com/2012/04/09/facebook-to-acquire-instagram-for-1-billion/

Coole, Diana, and Samantha Frost. 2010a. "Introducing the New Materialisms." In *New Materialisms: Ontology, Agency, and Politics*, edited by Diana Coole and Samantha Frost, 1–20. Durham, NC: Duke University Press.

Coole, Diana, and Samantha Frost, eds. 2010b. *New Materialisms: Ontology, Agency, and Politics*. Durham, NC: Duke University Press.

Cooley, Heidi Rae. 2004. "It's All about the *Fit*: The Hand, the Mobile Screenic Device and Tactile Vision." *Journal of Visual Culture* 3 (2): 133–155.

Corona, Alice. 2015. "Behind the NSA: Details and Images on Almost 300 Patents Filed by the National Security Agency." *Medium*, August 11. https://medium.com/silk-stories/behind-the-nsa-e0bf2c3a40c0

Corrigan, Thomas F. 2018. "Making Implicit Methods Explicit: Trade Press Analysis in the Political Economy of Communication." *International Journal of Communication* 12: 2751–2772.

Coscarelli, Joe. 2013. "Tweeting Every US Drone Strike Is Taking Way Longer Than Expected." *New York Magazine*, January 8. http://nymag.com/daily/intelligencer/2013/01/josh-begley-interview-on-dronestream-twitter-account.html

Costa, Elisabetta. 2018. "Affordances-in-practice: An Ethnographic Critique of Social Media Logic and Context Collapse." *New Media & Society*. https://doi.org/10.1177/1461444818756290

Coté, Mark. 2014. "Data Motility: The Materiality of Big Social Data." *Cultural Studies Review* 20 (1): 121–149. http://epress.lib.uts.edu.au/journals/index.php/csrj/article/view/3832/3962

Couldry, Nick. 2012. *Media, Society, World: Social Theory and Digital Media Practice*. Cambridge: Polity Press.

Coyne, Richard. 2010. *The Tuning of Place: Sociable Spaces and Pervasive Digital Media*. Cambridge, MA: MIT Press.

Crain, Matthew. 2018. "The Limits of Transparency: Data Brokers and Commodification." *New Media & Society* 20 (1): 88–104.

Cramer, Henriette, Mattias Rost, and Lars Erik Holmquist. 2011. "Performing a Check-in: Emerging Practices, Norms and 'Conflicts' in Location-sharing Using Foursquare." MobileHCI 2011, Stockholm, Sweden.

Crampton, Jeremy W. 2009. "Cartography: Maps 2.0." *Progress in Human Geography* 33 (1): 91–100.

Crampton, Jeremy W. 2015. "Collect It All: National Security, Big Data and Governance." *GeoJournal* 80: 519–531.

Crawford, Alice. 2008. "Taking Social Software to the Streets: Mobile Cocooning and the (An-)Erotic City." *Journal of Urban Technology* 15 (3): 79–97.

Crawford, Alice, and Gerard Goggin. 2009. "Geomobile Web: Locative Technologies and Mobile Media." *Australian Journal of Communication* 36 (1): 97–109.

Crawford, Kate. 2012. "Four Ways of Listening with an iPhone: From Sound and Network Listening to Biometric Data and Geolocative Tracking." In *Studying Mobile Media: Cultural Technologies, Mobile Communication, and the iPhone*, edited by Larissa Hjorth, Jean Burgess and Ingrid Richardson, 213–228. New York: Routledge.

Crook, Jordan. 2012. "Instagram Will Share User Data with Facebook According to Its New Privacy Policy." *TechCrunch*, December 17. http://techcrunch.com/2012/12/17/instagram-will-share-users-data-with-facebook-according-to-its-new-privacy-policy/

Crook, Jordan. 2017a. "Foursquare Launches Pilgrim SDK to Let Developers Leverage Location." *TechCrunch*, March 1. https://techcrunch.com/2017/03/01/foursquare-launches-pilgrim-sdk-to-let-developers-leverage-location/

Crook, Jordan. 2017b. "Swarm Kills the Clutter, Focuses on Logging Location." *TechCrunch*, August 8. https://techcrunch.com/2017/08/08/swarm-kills-the-clutter-focuses-on-logging-location/

Crook, Jordan. 2018. "Foursquare Is Finally Proving Its (Dollar) Value." *TechCrunch*, January 20. https://techcrunch.com/2018/01/19/foursquare-is-finally-proving-its-dollar-value/

Cuijpers, Colette, and Martin Pekárek. 2011. "The Regulation of Location-Based Services: Challenges to the European Union Data Protection Regime." *Journal of Location Based Services* 5 (3–4): 223–241.

Cunningham, Chris. 2017. "How Location Data Will Create the Cities of the Future." *GeoMarketing*, November 30. http://www.geomarketing.com/how-location-data-will-create-the-cities-of-the-future

Dale, Joshua Paul, Joyce Goggin, Julia Leyda, Anthony P. McIntyre, and Diane Negra, eds. 2017. *The Aesthetics and Affects of Cuteness*. New York: Routledge.

Daley, Angela, S. Kate Devitt, and Monique Mann, eds. 2019. *Good Data*. Amsterdam: Institute for Network Cultures.

Dash, Anil. 2013. "On Location with Foursquare." *Anil Dash: A Blog about Making Culture*, August 6. http://dashes.com/anil/2013/08/on-location-with-foursquare.html

Davey, Nicholas. 2012. "Inbetween Word and Image: Philosophical Hermeneutics, Aesthetics and the Inescapable Heritage of Kant." In *Critical Communities and Aesthetic Practices: Dialogues with Tony O'Connor on Society, Art and Friendship*, edited by Francis Halsall, Julia Jansen and Sinéad Murphy, 23–36. Dordrecht: Springer.

Davis, Jenny L., and Nathan Jurgenson. 2013. "Context Collapse: A Literature Review." *Cyborgology*, January 10. https://thesocietypages.org/cyborgology/2013/01/10/context-collapse-a-literature-review/

Davis, Jenny L., and Nathan Jurgenson. 2014. "Context Collapse: Theorizing Context Collusions and Collisions." *Information, Communication & Society* 17 (4): 476–485.

de Solier, Isabelle. 2013. *Food and the Self: Consumption, Production and Material Culture*. New York: Bloomsbury.

de Souza e Silva, Adriana. 2013. "Location-Aware Mobile Technologies: Historical, Social and Spatial Approaches." *Mobile Media & Communication* 1 (1): 116–121.

de Souza e Silva, Adriana, and Jordan Frith. 2010. "Locational Privacy in Public Spaces: Media Discourses on Location-Aware Mobile Technologies." *Communication, Culture & Critique* 3: 503–525.

de Souza e Silva, Adriana, and Jordan Frith. 2012. *Mobile Interfaces in Public Spaces: Locational Privacy, Control, and Urban Sociability*. New York: Routledge.

Debatin, Bernhard. 2011. "Ethics, Privacy, and Self-Restraint in Social Networking." In *Privacy Online: Perspectives on Privacy and Self-Disclosure in the Social Web*, edited by Sabine Trepte and Leonard Reinecke, 47–60. Heidelberg: Springer-Verlag.

Debatin, Bernhard, Jennette P. Lovejoy, Ann-Kathrin Horn, and Brittany N. Hughes. 2009. "Facebook and Online Privacy: Attitudes, Behaviors, and Unintended Consequences." *Journal of Computer-Mediated Communication* 15 (1): 83–108.

Dediu, Horace. 2015. "Where Are Maps Going?" *Asymco.com*, June 15. http://www.asymco.com/2015/06/15/where-are-maps-going/

Dell, Simon. 2011. "Why Do We Check In?" *Marketing Magazine*, November: 30–31.

Delo, Cotton. 2014. "How Foursquare Uses Location Data to Target Ads on PCs, Phones: Social Network Looks to Turn Location Data into Gold." *Ad Age*, February 27. http://adage.com/article/digital/foursquare-location-data-target-ads-web/291883/

Dena, Christy. 2008. "Emerging Participatory Culture Practices: Player-Created Tiers in Alternate Reality Games." *Convergence: The International Journal of Research into New Media Technologies* 14 (1): 41–57.

Dencik, Lina, Arne Hintz, and Jonathan Cable. 2016. "Towards Data Justice? The Ambiguity of Anti-surveillance Resistance in Political Activism." *Big Data & Society* (July–December). https://doi.org/10.1177/2053951716679678

Deumert, Ana. 2014. "The Performance of a Ludic Self on Social Network(ing) Sites." In *The Language of Social Media: Identity and Community on the Internet*, edited by Philip Seargeant and Caroline Tagg, 23–45. Houndmills, Basingstoke, Hampshire: Palgrave Macmillan.

Dienlin, Tobias, and Miriam J. Metzger. 2016. "An Extended Privacy Calculus Model for SNSs: Analyzing Self-disclosure and Self-withdrawal in a Representative U.S. Sample." *Journal of Computer-Mediated Communication* 21: 368–383.

Dieter, Michael. 2015. "Locative Aesthetics and the Actor-Network." In *Locative Media*, edited by Rowan Wilken and Gerard Goggin, 224–236. New York: Routledge.

Diffie, Whitfield, and Susan Landau. 2007. "The Export of Cryptography in the 20th Century and the 21st." In *The History of Information Security: A Comprehensive Handbook*, edited by Karl De Leeuw and Jan Bergstra, 725–736. Amsterdam: Elsevier.

Dobson, Mike. 2011. "Google, Navteq and Map Compilation." *Exploring Local*, July 6. http://blog.telemapics.com/?p=380

Dobson, Mike. 2012a. "Apple and Mapping?" *Exploring Local*, June 13. http://blog.telemapics.com/?p=386

Dobson, Mike. 2012b. "Google Maps Announces a 400 Year Advantage over Apple Maps." *Exploring Local*, September 20. http://blog.telemapics.com/?p=399

Dobson, Mike. 2012c. "*The Atlantic Magazine* Reveal How Google Build Its Maps—Not." *Exploring Local*, September 19. http://blog.telemapics.com/?p=394

Dobson, Mike. 2014. "Google Maps and Business Listings—Better, but Not Quite There." *Exploring Local*, December 16. http://blog.telemapics.com/?p=528

Dobson, Mike. 2015a. "Google Maps Stumble Badly—Crowdsourcing the Problem." *Exploring Local*, May 25. http://blog.telemapics.com/?p=577

Dobson, Mike. 2015b. "Can Anyone Stay on Top of the Online Mapping Hill?" *Exploring Local*, July 19. http://blog.telemapics.com/?p=584

Dobyns, Clifford. 2015. "Trace That Call." *Days Gone By*, September 20. http://daysgoneby.me/trace-call-1950s-style-took-little-longer/

Dodd, Weldon. 2010. "Report: Apple Acquires Poly9 Mapping Company." *Gigaom*, July 14. https://gigaom.com/2010/07/14/apple-acquires-poly9-mapping-company/

Dodge, Martin, and Rob Kitchin. 2007. "'Outlines of a World Coming into Existence': Pervasive Computing and the Ethics of Forgetting." *Environment and Planning B: Planning and Design* 34: 431–445.

Dolphijn, Rick, and Iris van der Tuin, eds. 2012. *New Materialism: Interviews and Cartographies*. Ann Arbor, MI: Open Humanities Press.

Dorling, Philip. 2013. "Pine Gap Drives US Drone Kills." *The Sydney Morning Herald*, July 21. http://www.smh.com.au/national/pine-gap-drives-us-drone-kills-20130720-2qbsa.html

Dourish, Paul. 2015. "Protocols, Packets, and Proximity: The Materiality of Internet Routing." In *Signal Traffic: Critical Studies of Media Infrastructures*, edited by Lisa Parks and Nicole Starosielski, 183–204. Urbana: University of Illinois.

du Gay, Paul, and Michael Pryke. 2002. "Cultural Economy: An Introduction." In *Cultural Economy: Cultural Analysis and Commercial Life*, edited by Paul du Gay and Michael Pryke, 1–19. London: Sage.

Dua, Kunal. 2017. "WhatsApp Live Location Sharing Launched: Here's How It Works." *Gadgets360*, October 18. https://gadgets.ndtv.com/apps/news/whatsapp-live-location-sharing-feature-tracking-apk-android-how-to-get-it-1764236

Duarte, Fernanda. 2015. "Rerouting Borders: Politics of Mobility and the Transborder Immigrant Tool." In *Mobility and Locative Media: Mobile Communication in Hybrid Spaces*, edited by Adriana de Souza e Silva and Mimi Sheller, 65–81. New York: Routledge.

DuVander, Adam. 2011. "Google Maps Cost: How Many Developers Will Have to Pay?" *ProgrammableWeb*, October 27. https://www.programmableweb.com/news/google-maps-cost-how-many-developers-will-have-to-pay/2011/10/27

DuVander, Adam. 2012. "SimpleGeo APIs Closed, but Places Data Is Open." *ProgrammableWeb*, January 12. https://www.programmableweb.com/news/simplegeo-apis-closed-places-data-open/2012/01/12

Economist. 2016. "High-definition Maps: The Autonomous Car's Reality Check." *The Economist*, April 16. https://www.economist.com/science-and-technology/2016/04/16/the-autonomous-cars-reality-check

Edmond, Rod. 1997. *Representing the South Pacific: Colonial Discourse from Cook to Gauguin*. Cambridge: Cambridge University Press.

Edwards, Benj. 2015. "Who Needs GPS? The Forgotten Story of Etak's Amazing 1985 Car Navigation System." *Fast Company*, June 26. https://www.fastcompany.com/3047828/who-needs-gps-the-forgotten-story-of-etaks-amazing-1985-car-navigation-system

Edwards, Jim. 2013. "Facebook Is Holding an Axe over the Neck of Foursquare." *Business Insider Australia*, October 20. http://www.businessinsider.com.au/facebook-is-holding-an-ax-over-the-neck-of-foursquare-2013-10

Edwards, Paul N. 2003. "Infrastructure and Modernity: Force, Time, and Social Organization in the History of Sociotechnical Systems." In *Modernity and Technology*, edited by Thomas J. Misa, Philip Brey, and Andrew Feenberg, 185–225. Cambridge, MA: MIT Press.

Elbaz, Gil. 2012. "Data Markets: The Emerging Data Economy." *TechCrunch*, October 1. https://techcrunch.com/2012/09/30/data-markets-the-emerging-data-economy/

Eldon, Eric. 2012. "Glancee: A Nice-Guy Ambient Social Location App for Normal People." *TechCrunch*, February 9. http://techcrunch.com/2012/02/09/glanceelocationapp/

Elliott, Anthony. 2014. *Concepts of the Self*. 3rd ed. Cambridge: Polity.

Elliott, Anthony. 2016. *Identity Troubles: An Introduction*. London: Routledge.

Ellison, P. Michele. 2012. "Notice of Apparent Liability for Forfeiture, File No. EB-10-JH-4055, NAL/Acct. No. 201232080020, FRNs 0010119691, 0014720239." Federal Communications Commission, April 13. https://www.wired.com/images_blogs/threatlevel/2012/04/91652398-FCC-Report-on-Google-Street-View-personal-data-mining.pdf

Elmer, Greg. 2004. *Profiling Machines: Mapping the Personal Information Economy*. Cambridge, MA: MIT Press.

Elmer, Greg. 2010. "Locative Networking: Finding and Being Found." *Aether: The Journal of Media Geography* 5A: 18–26.

Elmer, Greg. 2015. "Going Public on Social Media." *Social Media + Society* (April-June). https://doi.org/10.1177/2056305115580341

Elwood, Sarah, and Agnieszka Leszczynski. 2011. "Privacy, Reconsidered: New Representations, Data Practices, and the Geoweb." *Geoforum* 42: 6–15.

Erickson, Andrew S. 2012. "Microsatellites: A Bellwether of Chinese Aerospace Progress?" In *Down to Earth: Satellite Technologies, Industries, and Cultures*, edited by Lisa Parks and James Schwoch, 254–279. New Brunswick, NJ: Rutgers University Press.

Esayas, Samson Y. 2017. "The Idea of 'Emergent Properties' in Data Privacy: Towards a Holistic Approach." *International Journal of Law and Information Technology* 25: 139–178.

Esposito, Richard, Matthew Cole, Mark Schone, and Glenn Greenwald. 2014. "Snowden Docs Reveal British Spies Snooped on YouTube and Facebook." *NBC News*, January 27. http://www.nbcnews.com/news/investigations/exclusive-snowden-docs-reveal-uk-spies-snooped-youtube-facebook-n17381

Etherington, Darrell. 2018. "Intel's Mobileye Will Have 2 Million Cars on Roads Building HD Maps in 2018." *TechCrunch*, January 9. https://techcrunch.com/2018/01/08/intels-mobileye-will-have-2-million-cars-on-roads-building-hd-maps-in-2018/

Eustace, Alan. 2010a. "Creating Stronger Privacy Controls Inside Google." *Google Official Blog*, October 22. https://googleblog.blogspot.com/2010/10/creating-stronger-privacy-controls.html

Eustace, Alan. 2010b. "WiFi Data Collection: An Update." *Google Official Blog*, May 14. https://googleblog.blogspot.com/2010/05/wifi-data-collection-update.html

Evans, Benedict. 2013. "Foursquare Traction." *Benedict Evans*, October 9. http://ben-evans.com/benedictevans/2013/10/9/foursquare-traction

Evans, Benedict. 2014. "WhatsApp and $19bn." *Benedict Evans*, February 19. http://ben-evans.com/benedictevans/2014/2/19/whatsapp-and-19bn

Evans, Benedict (@BenedictEvans). 2015. "Reminder: Google Maps on iOS Only Has 100m or So Monthly Active Users, Out of Perhaps 500m in Use." Twitter, October 1, 2015, 10:57 a.m. https://twitter.com/BenedictEvans/status/649387577661722624

Evans, David S., Andrei Hagiu, and Richard Schmalensee. 2006. *Invisible Engines: How Software Platforms Drive Innovation and Transform Industries*. Cambridge, MA: MIT Press.

Evans, Leighton. 2014. "Maps as Deep: Reading the Code of Location-Based Social Networks." *IEEE Technology and Society Magazine* (Spring): 73–80.

Evans, Leighton. 2015. *Locative Social Media: Place in the Digital Age*. Houndmills, Basingstoke, Hampshire: Palgrave Macmillan.

Evans, Leighton, and Michael Saker. 2017. *Location-Based Social Media: Space, Time and Identity*. Cham, Switzerland: Palgrave Macmillan.

Evans, Sandra K., Katy E. Pearce, Jessica Vitak, and Jeffrey W. Treem. 2017. "Explicating Affordances: A Conceptual Framework for Understanding Affordances in Communication Research." *Journal of Computer-Mediated Communication* 22: 35–52.

Eyres, Patrick. 1986. "Despatches from the Little Spartan War." *New Arcadian Journal* 23.

Facebook. 2012. "Facebook Reports Second Quarter 2012 Results." *Facebook Investor Relations*, July 26. https://investor.fb.com/investor-news/press-release-details/2012/Facebook-Reports-Second-Quarter-2012-Results/default.aspx

Facebook. 2019. "Facebook Reports Fourth Quarter and Full Year 2018 Results." *Facebook Investor Relations*, January 30. https://s21.q4cdn.com/399680738/files/doc_financials/2018/Q4/Q4-2018-Earnings-Release.pdf

"Facebook Reveals Requests for User Data in Response to PRISM Scandal." 2013. *The Australian*, June 15. https://www.theaustralian.com.au/business/technology/facebook-reveals-requests-for-user-data-in-response-to-prism-scandal/news-story/9ea1c4a987db07760d4e7a47d72d763d

Falkheimer, Jesper, and André Jansson, eds. 2006. *Geographies of Communication: The Spatial Turn in Media Studies*. Göteborg: Nordicom.

Farman, Jason. 2012. *Mobile Interface Theory*. New York: Routledge.

Farman, Jason. 2014a. "Site-Specificity, Pervasive Computing, and the Reading Interface." In *The Mobile Self: Narrative Practices with Locative Technologies*, edited by Jason Farman, 3–16. New York: Routledge.

Farman, Jason, ed. 2014b. *The Mobile Story: Narrative Practices with Locative Technologies*. New York: Routledge.

Farman, Jason. 2015a. "Map Interfaces and Production of Locative Media Space." In *Locative Media*, edited by Rowan Wilken and Gerard Goggin, 83–93. New York: Routledge.

Farman, Jason. 2015b. "Objects as Audience: Phenomenologies of Vibrant Materiality in Locative Art." *Leonardo Electronic Almanac* 21 (1): 196–209.

Farman, Jason. 2015c. "The Forgotten Kaleidoscope Craze in Victorian England." *Atlas Obscura*, November 9. https://www.atlasobscura.com/articles/the-forgotten-kaleidoscope-craze-in-victorian-england

Farman, Jason. 2015d. "The Materiality of Locative Media: On the Invisible Infrastructure of Mobile Networks." In *Theories of the Mobile Internet: Materialities and Imaginaries*, edited by Andrew Herman, Jan Hadlaw, and Thom Swiss, 45–59. New York: Routledge.

Farman, Jason. 2018a. *Delayed Response: The Art of Waiting from the Ancient to the Instant World*. New Haven, CT: Yale University Press.

Farman, Jason. 2018b. "Invisible and Instantaneous: Geographies of Media Infrastructure from Pneumatic Tubes to Fiber Optics." *Media Theory* 2 (1). http://mediatheoryjournal.org/jason-farman-invisible-and-instantaneous/

Fast, Karin, André Jansson, Johan Lindell, Linda Ryan Bengtsson, and Mekonnen Tesfahuney, eds. 2018. *Geomedia Studies: Spaces and Mobilities in Mediatized Worlds*. New York: Routledge.

Fenton, Natalie. 2007. "Bridging the Mythical Divide: Political Economy and Cultural Studies Approaches to the Analysis of the Media." In *Media Studies: Key Issues and Debates*, edited by Eoin Devereux, 7–31. London: SAGE.

Fernback, Jan, and Zizi Papacharissi. 2007. "Online Privacy as Legal Safeguard: The Relationship among Consumer, Online Portal, and Privacy Policies." *New Media & Society* 9 (5): 715–734.

Ferrari, Laura, and Marco Mamei. 2011. "Discovering Daily Routines from Google Latitude with Topic Models." 8th IEEE Workshop on Context Modeling and Reasoning, Seattle, WA, March 21–25, 2011, 432–437. https://ieeexplore.ieee.org/stamp/stamp.jsp?arnumber=5766928

Fingas, Jon. 2018. "Foursquare Puts Check-in Data to Greater Use in Apps like Snapchat." *Engadget*, May 30. https://www.engadget.com/2018/05/30/foursquare-mapbox-deal/

Fiore-Gartland, Brittany, and Gina Neff. 2015. "Communication, Mediation, and the Expectations of Data: Data Valences across Health and Wellness Communities." *International Journal of Communication* 9: 1466–1484.

Fisher, Steven. 2011. "How the iPhone GPS Differs from a Standalone Navigation GPS." *tewha.net*, October 29. http://tewha.net/2011/10/how-the-iphone-gps-differs-from-a-standalone-navigation-gps/

Fleet Owner. 2012. "Wingcast Project Dissolved." *FleetOwner*, July 1. http://fleetowner.com/mag/fleet_wingcast_project_dissolved

Fleischer, Peter. 2010. "Data Collected by Google Cars." *Google Europe Blog*, April 27. https://europe.googleblog.com/2010/04/data-collected-by-google-cars.html

Fleishman, Glenn. 2011. "How the iPhone Knows Where You Are." *Macworld*, April 28. http://www.macworld.com/article/1159528/smartphones/how-iphone-location-works.html

Flew, Terry. 2009. "The Cultural Economy Moment?" *Cultural Science* 2 (1). http://cultural-science.org/journal/index.php/culturalscience/article/view/28/55

Flynn, Kerry. 2017. "Foursquare Just Unveiled Its Biggest Project Yet." *Mashable*, March 2. http://mashable.com/2017/03/01/foursquare-mobile-data-pilgrim/#qk0LJsf6rkqu

Forlano, Laura. 2013. "Making Waves: Urban Technology and the Co-production of Place." *First Monday* 18 (11). http://firstmonday.org/ojs/index.php/fm/article/view/4968/3797w

Forrest, Brady. 2007. "Poly9 Free Earth." *O'Reilly Radar*, May 10. http://radar.oreilly.com/2007/05/poly9-free-earth.html

Fortunati, Leopoldina. 2005. "Mobile Telephone and the Presentation of Self." In *Mobile Communications: Re-negotiation of the Social Sphere*, edited by Rich Ling and Per E. Pedersen, 203–218. London: Springer.

Fougner, Jon. 2011. "Introducing Deals." *The Facebook Blog*, February 1. https://blog.facebook.com/blog.php?post=446183422130

Foursquare. 2018. "Driving with Data: How Location Tech Is Reinventing the Auto Industry." *Medium*, July 21. https://medium.com/foursquare-direct/driving-with-data-how-location-tech-is-reinventing-the-auto-industry-3e67bdb6a8cf

Foursquare Blog. 2013a. "Big News: Today We're Opening Up Foursquare Ads to All Small Businesses around the World." *Foursquare Blog*, October 14. http://blog.foursquare.com/post/64014727485/big-news-today-were-opening-up-foursquare-ads-to-all-businesses-around-the-world/

Foursquare Blog. 2013b. "Manage a Business on Foursquare? Download Our New App to Easily Connect with Customers Right from Your Phone." *Foursquare Blog*, January 29. http://blog/foursquare.com/2013/01/29/manage-a-business-on-foursquare-download-our-new-app-to-easily-connect-with-customers-right-from-your-phone/

Foursquare Filling. 2013. "Foursquare Filling in Venue Details by Asking Users within the App." *About Foursquare*, August 8. http://aboutfoursquare.com/foursquare-filling-in-venue-details-by-asking-users-within-the-app/

Foust, Jeff. 2017. "Is the Earth-Observation Industry Consolidating, or Just Evolving?" *SpaceNews*, June 14. http://spacenews.com/is-the-earth-observation-industry-consolidating-or-just-evolving/

Franceschi-Bicchierai, Lorenzo. 2014. "After 5 Rejections, Apple Accepts App That Tracks US Drone Strikes." *Mashable*, February 8. http://mashable.com/2014/02/07/apple-app-tracks-drone-strikes/#S03T4.4QwmqN

Francica, Joe. 2002. "Yahoo Boots MapQuest." *Directions Magazine*, March 14. https://www.directionsmag.com/article/3940

Freund, Paul A. 1975. "Address to the American Law Institute." Quoted in the Proceedings of the 52nd Annual Meeting of the American Law Institute, 574–575.

Fried, Charles. 1968. "Privacy." *The Yale Law Journal* 77 (3): 475–493.

Fried, Ina. 2013. "Apple Did Indeed Acquire BroadMap and Catch Earlier this Year." *All Things D*, December 23. http://allthingsd.com/20131223/apple-did-indeed-acquire-broadmap-and-catch-earlier-this-year/

Frith, Jordan. 2013. "Turning Life into a Game: Foursquare, Gamification, and Personal Mobility." *Mobile Media & Communication* 1 (2): 248–262.

Frith, Jordan. 2014. "Communicating through Location: The Understood Meaning of the Foursquare Check-in." *Journal of Computer-Mediated Communication* 19: 890–905.

Frith, Jordan. 2015a. "Communicating behind the Scenes: A Primer on Radio Frequency Identification (RFID)." *Mobile Media & Communication* 3 (1): 91–105.

Frith, Jordan. 2015b. *Smartphones as Locative Media*. Cambridge: Polity.

Frith, Jordan. 2015c. "Writing Space: Examining the Potential of Location-based Composition." *Computers and Composition* 37: 44–54.

Frith, Jordan. 2017. "Invisibility through the Interface: The Social Consequences of Spatial Search." *Media Culture & Society* 39 (4): 536–551.

Frith, Jordan. 2019. *A Billion Little Pieces: RFID and Infrastructures of Identification*. Cambridge, MA: MIT Press.

Frith, Jordan, and Kati Fargo Ahern. 2015. "Make a Sound Garden Grow: Exploring the New Media Potential of Social Soundscaping." *Convergence: The International Journal of Research into New Media Technologies* 21 (4): 496–508.

Frith, Jordan, and Jason Kalin. 2016. "Here, I Used to Be: Mobile Media and Practices of Place-Based Digital Memory." *Space & Culture* 19 (1): 43–55.

Frith, Jordan, and Michael Saker. 2017. "Understanding Yik Yak: Location-based Sociability and the Communication of Place." *First Monday* 22 (10). http://firstmonday.org/ojs/index.php/fm/article/view/7442

Frommer, Dan. 2010. "Why Apple Bought Quattro Wireless and Is Getting into Advertising." *Business Insider Australia*, January 6. http://www.businessinsider.com.au/why-apple-bought-quattro-wireless-and-is-getting-into-mobile-advertising-2010-1

Fuchs, Christian. 2012. "The Political Economy of Privacy on Facebook." *Television & New Media* 13 (2): 139–159.

Fuller, Matthew. 2005. *Media Ecologies: Materialist Energies in Art and Technoculture.* Cambridge, MA: MIT Press.

Gale, Gary. 2012. "Map Wars; Are Apple's Maps Really That Bad?" *Mostly Maps*, September 24. https://www.vicchi.org/2012/09/24/map-wars-are-apples-maps-really-that-bad/

Gale, Gary. 2013. "The Ubiquitous Digital Map (Unabridged)." *Mostly Maps*, February 19. https://www.vicchi.org/2013/02/19/the-ubiquitous-digital-map-abridged/

Gale, Gary. 2015. "As Nokia Looks to Sell HERE Maps, the Map Wars Are Underway." *Mostly Maps*, April 13. http://www.vicchi.org/2015/04/13/as-nokia-looks-to-sell-here-maps-the-map-wars-are-underway/

Gallagher, Ryan. 2014. "How Secret Partners Expand NSA's Surveillance Dragnet." *The Intercept*, June 19. https://theintercept.com/2014/06/18/nsa-surveillance-secret-cable-partners-revealed-rampart-a/

Gallagher, Sean. 2014. "Where've You Been? Your Smartphone's Wi-Fi Is Telling Everyone." *Ars Technica*, November 6. https://arstechnica.com/information-technology/2014/11/where-have-you-been-your-smartphones-wi-fi-is-telling-everyone/

Galloway, Anne. 2013. "Affective Politics in Urban Computing and Locative Media." In *Throughout: Art and Culture Emerging with Ubiquitous Computing*, edited by Ulrik Ekman, 351–364. Cambridge, MA: MIT Press.

Garnham, Nicholas. 1995. "Political Economy and Cultural Studies: Reconciliation or Divorce?" *Critical Studies in Mass Communication* 12: 62–71.

Garrison, Jim. 2012. "Parsing a Text File or Data Stream." *StackOverflow*, December 18. https://stackoverflow.com/questions/13924111/what-does-this-mean-parsing-a-text-file-or-data-stream-and-does-it-apply-with

Gaver, William W. 1991. "Technology Affordances." Proceedings of the SGCHI Conference on Human Factors in Computing Systems, ACM, 79–84. http://dl.acm.org/citation.cfm?id=108856.

Gavison, Ruth. 1980. "Privacy and the Limits of Law." *The Yale Law Journal* 89 (3): 421–471.

Gazzard, Alison. 2011. "Location, Location, Location: Collecting Space and Place in Mobile Media." *Convergence: The International Journal of Research into New Media Technologies* 17 (4): 405–417.

Geens, Stefan. 2012. "Constraining Online Maps: The Case of South Korea." *Ogle Earth*, July 1. https://ogleearth.com/2012/07/constraining-online-maps-the-case-of-south-korea/

Geissler, Roger C. 2012. "Private Eyes Watching You: Google Street View and the Right to an Inviolate Personality." *Hastings Law Journal* 63: 897–926.

Gekker, Alex, Sam Hind, Sybille Lammes, Chris Perkins, and Clancy Wilmott. 2018. "Introduction: Mapping Times." In *Time for Mapping: Cartographic Temporalities*, edited by Sybille Lammes, Chris Perkins, Alex Gekker, Sam Hind, Clancy Wilmott, and Daniel Evans, 1–23. Manchester: Manchester University Press.

Geoawesomeness. 2016. "Geotrends 2016—What To Expect Across Geoindustries This Year?" *Geoawesomeness*, January 31. https://geoawesomeness.com/geotrends-2016-expect-across-geoindustries-year/

Gellman, Barton, and Greg Miller. 2013. "'Black Budget' Summary Details U.S. Spy Network's Successes, Failures and Objectives." *Washington Post*, August 29. https://www.washingtonpost.com/world/national-security/black-budget-summary-details-us-spy-networks-successes-failures-and-objectives/2013/08/29/7e57bb78-10ab-11e3-8cdd-bcdc09410972_story.html?utm_term=.5d51354f31a4

Gellman, Barton, and Laura Poitras. 2013. "US, British Intelligence Mining Data from Nine US Internet Companies in Broad Secret Program." *Washington Post*, June 7. https://www.washingtonpost.com/investigations/us-intelligence-mining-data-from-nine-us-internet-companies-in-broad-secret-program/2013/06/06/3a0c0da8-cebf-11e2-8845-d970ccb04497_story.html?utm_term=.c56229fc7ec9

Gellman, Barton, and Ashkan Soltani. 2013a. "NSA Infiltrates Links to Yahoo, Google Data Centers Worldwide, Snowden Documents Say." *Washington Post*, October 30. https://www.washingtonpost.com/world/national-security/nsa-infiltrates-links-to-yahoo-google-data-centers-worldwide-snowden-documents-say/2013/10/30/e51d661e-4166-11e3-8b74-d89d714ca4dd_story.html?utm_term=.a35b733c2c4f

Gellman, Barton, and Ashkan Soltani. 2013b. "NSA Tracking Cellphone Locations Worldwide, Snowden Documents Show." *Washington Post*, December 4. https://www.washingtonpost.com/world/national-security/nsa-tracking-cellphone-locations-worldwide-snowden-documents-show/2013/12/04/5492873a-5cf2-11e3-bc56-c6ca94801fac_story.html?utm_term=.5d62cc08ab60

Gellman, Barton, and Ashkan Soltani. 2014. "NSA Surveillance Program Reaches 'Into the Past' to Retrieve, Replay Phone Calls." *Washington Post*, March 18. https://www.washingtonpost.com/world/national-security/nsa-surveillance-program-reaches-into-the-past-to-retrieve-replay-phone-calls/2014/03/18/226d2646-ade9-11e3-a49e-76adc9210f19_story.html?utm_term=.7774d74a578b

Gerlach, Joe. 2018. "Nodes, Ways and Relations." In *Time for Mapping: Cartographic Temporalities*, edited by Sybille Lammes, Chris Perkins, Alex Gekker, Sam Hind, Clancy Wilmott and Daniel Evans, 27–49. Manchester: Manchester University Press.

Gerlitz, Carolin, and Anne Helmond. 2013. "The Like Economy: Social Buttons and the Data-Intensive Web." *New Media & Society* 15 (8): 1348–1365.

Geron, Tomio. 2012. "Facebook Takes on Foursquare and Yelp with Nearby." *Forbes*, December 17. http://www.forbes.com/sites/tomiogeron/2012/12/17/facebook-takes-on-foursquare-and-yelp-with-location/

Gibson, Chris, and Lily Kong. 2005. "Cultural Economy: A Critical Review." *Progress in Human Geography* 29 (5): 541–561.

Gibson, James J. 2015. *The Ecological Approach to Visual Perception*. New York: Psychology Press.

Gidda, Mirren. 2013. "Edward Snowden and the NSA Files—Timeline." *The Guardian*, August 22. https://www.theguardian.com/world/2013/jun/23/edward-snowden-nsa-files-timeline

Giddens, Anthony. 1991. *Modernity and Self-identity: Self and Society in the Late Modern Age*. Stanford, CA: Stanford University Press.

Gillespie, Tarleton. 2010. "The Politics of 'Platforms.'" *New Media & Society* 12 (3): 347–364.

Gillespie, Tarleton, Pablo J. Boczkowski, and Kirsten A. Foot, eds. 2014. *Media Technologies: Essays on Communication, Materiality, and Society*. Cambridge, MA: MIT Press.

Girardin, Fabien, Andrea Vaccari, Alexandre Gerber, Assaf Biderman, and Carlo Ratti. 2009. "Towards Estimating the Presence of Visitors from the Aggregate Mobile Phone Network Activity They Generate." 11th International Conference on Computers in Urban Planning and Urban Management (CUPUM), Hong Kong, June 16–18. http://www.girardin.org/fabien/publications/girardin_et_al_cupum09_final_pre_editing.pdf

Gitelman, Lisa. 2006. *Always Already New: Media, History, and the Data Culture*. Cambridge, MA: MIT Press.

Gitelman, Lisa, and Virginia Jackson. 2013. "Introduction." In *"Raw Data" Is an Oxymoron*, edited by Lisa Gitelman, 1–14. Cambridge, MA: MIT Press.

Glanz, James, Jeff Larson, and Andrew W. Lehren. 2014. "Spy Agencies Tap Data Streaming from Phone Apps." *New York Times*, January 27. https://www.nytimes.com/2014/01/28/world/spy-agencies-scour-phone-apps-for-personal-data.html?mcubz=0

Gobry, Pascal-Emmanuel. 2011. "Foursquare Gets 3 Million Check-ins per Day, Signed Up 500,000 Merchants." *Business Insider*, August 2. www.businessinsider.com/foursquare-dennis-crowley-interview-2011-8

Gobry, Pascal-Emmanuel. 2012. "Foursquare's Revenue Model Sharpens into Focus." *Business Insider*, May 9. http://articles.businessinsider.com/2012-05-09/research/31634849_1_foursquare-business-model-yelp

Godlewska, Anne, and Jason Grek Martin. 2011. "Map." In *The SAGE Handbook of Geographical Knowledge*, edited by John A. Agnew and David N. Livingstone, 357–367. London: SAGE.

Goffman, Erving. 1971 [1959]. *The Presentation of Self in Everyday Life*. London: Penguin.

Goggin, Gerard. 2006. *Cell Phone Culture*. London: Routledge.

Goggin, Gerard. 2011. *Global Mobile Media*. London: Routledge.

Goggin, Gerard. 2012. "List Media: The Telephone Directory and the Arranging of Names." *M/C Journal* 15 (5). http://journal.media-culture.org.au/index.php/mcjournal/article/view/556

Goggin, Gerard. 2014. "Facebook's Mobile Career." *New Media & Society* 16 (7): 1068–1086.

Goggin, Gerard, and Mark McLelland. 2009. "Internationalizing Internet Studies—Beyond Anglophone Paradigms." In *Internationalizing Internet Studies: Beyond Anglophone Paradigms*, edited by Gerard Goggin and Mark McLelland, 3–17. New York: Routledge.

Goggin, Gerard, Ariadne Vromen, Kimberlee Weatherall, Fiona Martin, Adele Webb, Lucy Sunman, and Francesco Bailo. 2017. *Digital Rights in Australia*. Sydney: University of Sydney.

Golding, Peter, and Graham Murdock. 2000. "Culture, Communications and Political Economy." In *Mass Media and Society*, edited by James Curran and Michael Gurevitch, 15–30. London: Arnold.

Goldman, David. 2012. "Foursquare CEO: 'Not Just Check-ins and Badges.'" *CNN Money*, February 29. https://money.cnn.com/2012/02/29/technology/foursquare_ceo/index.htm

Goldsmith, Ben. 2014. "The Smartphone App Economy and App Ecosystem." In *The Routledge Companion to Mobile Media*, edited by Gerard Goggin and Larissa Hjorth, 171–180. New York: Routledge.

Gompers, Paul, and Josh Lerner. 2004. *The Venture Capital Cycle*. 2nd ed. Cambridge, MA: MIT Press.

Google. 2012. "Google Inc. Announces Third Quarter 2012 Results." *Google Inc.*, October 18. https://investor.google.com/earnings/2012/Q3_google_earnings.html

Gopinath, Sumanth. 2013. *The Ringtone Dialectic: Economy and Cultural Form*. Cambridge, MA: MIT Press.

Gordon, Eric, and Adriana de Souza e Silva. 2011. *Net Locality: Why Location Matters in a Networked World*. Chichester, West Sussex: Wiley-Blackwell.

Gormley, Ken. 1992. "One Hundred Years of Privacy." *Wisconsin Law Review* 1992 (5): 1335–1441.

Gray, Mary L. 2015. "Putting Social Media in Its Place: A Curatorial Theory for Media's Noisy Social Worlds." *Social Media + Society* (April-June). https://doi.org/10.1177/2056305115578683

Green, Nicola. 2009. "Mobility, Memory, and Identity." In *Mobile Technologies: From Telecommunications to Media*, edited by Gerard Goggin and Larissa Hjorth, 266–281. New York: Routledge.

Greenberg, Joshua. 2008. *From Betamax to Blockbuster: Video Stores and the Invention of Movies on Video*. Cambridge, MA: MIT Press.

Greene, Dan, and Katie Shilton. 2018. "Platform Privacies: Governance, Collaboration, and the Different Meanings of 'Privacy' in iOS and Android Development." *New Media & Society* 20 (4): 1640–1657.

Greengard, Samuel. 2015. *The Internet of Things*. Cambridge, MA: MIT Press.

Greenwald, Glenn. 2014. *No Place to Hide: Edward Snowden, the NSA, and the U.S. Surveillance State*. New York: Metropolitan Books.

Greenwald, Glenn, and Ewen MacAskill. 2013. "Boundless Informant: The NSA's Secret Tool to Track Global Surveillance Data." *The Guardian*, June 11. https://www.theguardian.com/world/2013/jun/08/nsa-boundless-informant-global-datamining

Greenwald, Glenn, Laura Poitras, and Jeremy Scahill. 2014. "Welcome to *The Intercept*." *The Intercept*, February 10. https://theintercept.com/2014/02/10/welcome-intercept

Griffiths, Tom. 2016. *The Art of Time Travel: Historians and Their Craft*. Carlton, Victoria: Black Inc.

Gross, Ralph, and Alessandro Acquisti. 2005. "Information Revelation and Privacy in Online Social Networks." WPES'05, November 7, 2005, Alexandria, VA, 71–80.

Grossberg, Larry. 1995. "Cultural Studies vs Political Economy: Is Anyone Else Bored with this Debate?" *Critical Studies in Mass Communication* 12: 72–81.

Grothaus, Michael. 2013. "An In-depth Comparison between iOS Map Frameworks: Apple MapKit vs. Google Maps SDK." *Fast Company*, March 18. http://www.fastcompany.com/3006725/open-company/depth-comparison-between-ios-map-frameworks-apple-mapkit-vs-google-maps-sdk

Guardian. 2015. "The Counted: People Killed by Police in the US." *The Guardian*,
June 1. https://www.theguardian.com/us-news/ng-interactive/2015/jun/01/
the-counted-police-killings-us-database

Gundersen, Eric. 2018. "Mapbox Foursquare: Over 100 Million Places on the Map."
Mapbox, May 30. https://blog.mapbox.com/mapbox-%EF%B8%8F-foursquare-
over-100-million-places-on-the-map-3ccf8d260265

Gurman, Mark. 2011. "Apple Acquired Mind-blowing 3D Mapping Company C3
Technologies, Looking to Take iOS Maps to the Next Level." *9to5Mac*, October
29. http://9to5mac.com/2011/10/29/apple-acquired-mind-blowing-3d-
mapping-company-c3-technologies-looking-to-take-ios-maps-to-the-next-
level/

Gurman, Mark. 2013. "Apple Acquired Mapping Firm BroadMap's Talent, Location-
Infused Evernote Competitor Catch." *9to5Mac*, December 23. http://
9to5mac.com/2013/12/23/apple-likely-acquired-mapping-firm-broadmap-
location-infused-evernote-competitor-catch/

Gurman, Mark. 2015. "Mystery Solved: Apple Vans Gathering Next-Gen Maps Data."
9to5Mac, May 29. http://9to5mac.com/2015/05/29/mystery-solved-apple-
vans-gathering-next-gen-maps-data-grabbing-street-view-storefronts-3d-
images/

Gustafsson, Henrik. 2013. "Foresight, Hindsight and State Secrecy in the American
West: The Geopolitical Aesthetics of Trevor Paglen." *Journal of Visual Culture* 12
(1): 148–164.

Ha, Anthony. 2013. "Yep, Social Discovery Startup Sonar Is Dead (and Its CEO
Explains Why)." *TechCrunch*, September 17. https://techcrunch.com/2013/09/
17/rip-sonar/

Ha, Anthony. 2014. "In Its First Acquisition, YP Buys Mobile Ad Company Sense
Networks." *TechCrunch*, January 6. https://techcrunch.com/2014/01/06/
yp-acquires-sense-networks/

Hachman, Mark. 2010. "Skyhook Sues Google in Location Patent, Contract
Dispute." *PC Mag*, September 15. https://www.pcmag.com/article2/
0,2817,2369228,00.asp

Haislip, Alexander. 2011. *The Essentials of Venture Capital*. Hoboken, NJ: John Wiley
& Sons.

Haleqoua, Germaine R., Alex Leavitt, and Mary L. Gray. 2016. "Jumping for Fun?
Negotiating Mobility and the Geopolitics of Foursquare." *Social Media + Society*.
https://doi.org/10.1177/2056305116665859/

Halliday, Josh. 2010a. "Google's South Korean Office Raided." *The Guardian*,
August 10. https://www.theguardian.com/technology/2010/aug/10/
google-street-view-seoul-police-raid

Halliday, Josh. 2010b. "Google Street View Broke Canada's Privacy Law with Wi-
Fi Capture." *The Guardian*, October 20. https://www.theguardian.com/
technology/2010/oct/19/google-street-view-privacy-canada

Hanke, John. 2007. "A Picture's Worth a Thousand Clicks." *Google Official Blog*, May
30. https://googleblog.blogspot.com/2007/05/pictures-worth-thousand-
clicks.html

Hardawar, Devindra. 2013. "Location Labs' Secret to Reaching 50M Installs: Give
Carriers Mobile Security Features to Call Their Own." *VentureBeat*, November
13. https://venturebeat.com/2013/11/13/location-labs-secret-to-reaching-
50m-installs-give-carriers-mobile-security-features-to-call-their-own/

Hardy, Jonathan. 2014. *Critical Political Economy of the Media*. London: Routledge.

Harris, Shane. 2014. "The NSA's Patents, in One Searchable Database."
  Foreign Policy, July 30. http://foreignpolicy.com/2014/07/30/
  the-nsas-patents-in-one-searchable-database/

Hartmann, Maren. 2011. "Mobile Privacy: Contexts." In *Privacy Online: Perspectives
  on Privacy and Self-Disclosure in the Social Web*, edited by Sabine Trepte and
  Leonard Reinecke, 191–203. Heidelberg: Springer-Verlag.

Hartzog, Woodrow, and Evan Selinger. 2013. "Obscurity: A Better Way to
  Think about Your Data than 'Privacy.'" *The Atlantic*, January 17. https://
  www.theatlantic.com/technology/archive/2013/01/obscurity-a-better-way-to-
  think-about-your-data-than-privacy/267283/

Hartzog, Woodrow, and Evan Selinger. 2015. "Surveillance as Loss of Obscurity."
  Washington and Lee Law Review 72 (3): 1343–1387.

Hayles, N. Katherine. 2013. "Radio-Frequency Identification: Human Agency and
  Meaning in Information-intensive Environments." In *Throughout: Art and
  Culture Emerging with Ubiquitous Computing*, edited by Ulrik Ekman, 503–528.
  Cambridge, MA: MIT Press.

Haythornthwaite, Caroline. 2005. "Social Networks and Internet Connectivity
  Effects." *Information, Communication & Society* 8 (2): 125–147.

He, Wenbo, Xue Liu, and Mai Ren. 2011. "Location Cheating: A Security
  Challenge to Location-based Social Network Services." 31st International
  Conference on Distributed Computing Systems (ICDCS), Minneapolis, MN,
  June 20–24.

Hearn, Gregory N., and Marcus Foth. 2007. "Communicative Ecologies: Editorial
  Preface." *The Electronic Journal of Communication / La Revue
  Électronique de Communication* 17 (1–2). http://www.cios.org/www/ejc/
  v17n12.htm#introduction

Heater, Brian. 2018. "Maps Walking Navigation Is Google's Most Compelling
  Use for AR Yet." *TechCrunch*, May 9. https://techcrunch.com/2018/05/08/
  maps-walking-navigation-is-googles-most-compelling-use-for-ar-yet/

Helmond, Anne. 2015. "The Platformization of the Web: Making Web Data Platform
  Ready." *Social Media + Society*. https://:doi.org/10.1177/2056305115603080

Hemment, Drew, ed. 2006. "Locative Media," special themed issue of *Leonardo
  Electronic Almanac* 14 (3). http://leoalmanac.org/journal/vol_14/lea_v14_n03-
  04/essays.asp

Henkin, Louis. 1974. "Privacy and Autonomy." *Columbia Law Review* 74
  (8): 1410–1433.

Hern, Alex. 2012. "Why Apple Had to Change Its Maps." *New Statesman*,
  September 27. https://www.newstatesman.com/blogs/sci-tech/2012/09/
  why-apple-had-change-its-maps

Hertig, Alyssa. 2018. "FOAM and the Dream to Map the World on
  Ethereum." *Coindesk*, June 7. https://www.coindesk.com/
  foam-dream-map-world-ethereum/

Highmore, Ben. 2011. *Ordinary Lives: Studies in the Everyday*. London: Routledge.

Hijleh, Aidah. 2012. "A Brief Guide to Facebook Single Sign On." *Soshable*, February
  16. http://soshable.com/facebook-single-sign-on/

Hill, Kashmir. 2010. "How Smartphone Apps Spy on Their Users." *Forbes*, December
  20. https://www.forbes.com/sites/kashmirhill/2010/12/20/how-smartphone-
  applications-spy-on-their-users/#2b9a6d292bd8

Hinsman, Abbey. 2014. "Undetected Media: Intelligence and the U-2 Spy Plane." *The
  Velvet Light Trap* 73 (Spring): 19–38.

Hjorth, Larissa. 2009. *Mobile Media in the Asia-Pacific: Gender and the Art of Being Mobile*. New York: Routledge.

Hjorth, Larissa. 2011. *Games and Gaming: An Introduction*. Oxford: Berg.

Hjorth, Larissa, Jean Burgess, and Ingrid Richardson, eds. 2012. *Studying Mobile Media: Cultural Technologies, Mobile Communication, and the iPhone*. New York: Routledge.

Hjorth, Larissa, Sarah Pink, and Heather Horst. 2018. "Being at Home with Privacy: Privacy and Mundane Intimacy through Same-Sex Locative Media Practices." *International Journal of Communication* 12: 1209–1227.

Hjorth, Larissa, and Ingrid Richardson. 2014. *Gaming in Social, Locative, & Mobile Media*. Houndmills, Basingstoke, Hampshire: Palgrave Macmillan.

Hoelzl, Ingrid, and Rémi Marie. 2014. "Google Street View: Navigating the Operative Image." *Visual Studies* 29 (3): 261–271.

Hogan, Bernie. 2010. "The Presentation of the Self in the Age of Social Media: Distinguishing Performances and Exhibitions Online." *Bulletin of Science, Technology & Society* 30 (6): 377–386.

Holt, Jennifer, and Patrick Vonderau. 2015. "'Where the Internet Lives': Data Centers as Cloud Infrastructure." In *Signal Traffic: Critical Studies of Media Infrastructures*, edited by Lisa Parks and Nicole Starosielski, 71–93. Urbana: University of Illinois Press.

Hong, Jason. 2012. "Analysis of Most Unexpected Permissions for Android Apps." *Jason Hong's Confabulations*, November 30. http://confabulator.blogspot.com.au/2012/11/analysis-of-top-10-most-unexpected.html

Horn, Leslie. 2010. "Pew: Four Percent of Adults Use Location-Based Services." *PCMag*, November 4. http://www.pcmag.com/article2/0,2817,2372092,00.asp

Horst, Heather. 2013. "The Infrastructures of Mobile Communication: Towards a Future Research Agenda." *Mobile Media & Communication* 1 (1): 147–152.

Hudson, Dale, and Patricia R. Zimmermann. 2015. *Thinking through Digital Media: Transnational Environments and Locative Places*. New York: Palgrave Macmillan.

Huffman, Stephen Mark, and Michael Henry Reifer. 2002. Method for Geolocating Logical Network Addresses. US Patent 2002/0087666 A1, filed December 29, 2000, and issued July 4, 2002.

Hughes, J. M., P. A. Michell, and W. S. Ramson, eds. 1992. *The Australian Concise Oxford Dictionary*. Melbourne: Oxford University Press.

Hulsey, Nathan. 2015. "Houses in Motion: An Overview of Gamification in the Context of Mobile Interfaces." In *Mobility and Locative Media: Mobile Communication in Hybrid Spaces*, edited by Adriana de Souza e Silva and Mimi Sheller, 149–164. New York: Routledge.

Hum, Noelle J., Perrin E. Chamberlin, Brittany L. Hambright, Anne C. Portwood, Amanda C. Schat, and Jennifer L. Bevan. 2011. "A Picture Is Worth a Thousand Words: A Content Analysis of Facebook Profile Photographs." *Computers in Human Behavior* 27: 1828–1833.

Humphreys, Lee. 2007. "Mobile Social Networks and Spatial Practice: A Case Study of Dodgeball." *Journal of Computer-Mediated Communication* 13 (1): 341–360.

Humphreys, Lee. 2010. "Mobile Social Networks and Urban Public Space." *New Media & Society* 12 (5): 763–778.

Humphreys, Lee. 2011. "Who's Watching Whom? A Study of Interactive Technology and Surveillance." *Journal of Communication* 61: 575–595.

Humphreys, Lee. 2012. "Connecting, Coordinating, Cataloguing: Communicative Practices on Mobile Social Networks." *Journal of Broadcasting & Electronic Media* 56 (4): 494–510.

Humphreys, Lee. 2018. *The Qualified Self: Social Media and the Accounting of Everyday Life*. Cambridge, MA: MIT Press.

Humphreys, Lee, and Tony Liao. 2013. "Foursquare and the Parochialization of Public Space." *First Monday* 18 (11). http://firstmonday.org/ojs/index.php/fm/article/view/4966

Humphreys, Lee, Thilo Von Pape, and Veronika Karnowski. 2013. "Evolving Mobile Media: Uses and Conceptualizations of the Mobile Internet." *Journal of Computer-Mediated Communication* 18: 491–507.

Humphreys, Lee, and Rowan Wilken. 2015. "Social Media, Small Business, and the Control of Information." *Information, Communication & Society* 18 (3): 295–309.

Hutcheon, Stephen. 2015. "The Untold Story about the Founding of Google Maps." *Medium*, February 10. https://medium.com/@lewgus/the-untold-story-about-the-founding-of-google-maps-e4a5430aec92

Hutchison Whampoa Limited. 2002. "Hutchison 3G UK Announces Revolutionary New Digital Mapping for Location-based Services." *Huchison Whampoa Limited*, September 18. http://www.hutchison-whampoa.com/en/media/press_each.php?id=1008

"Introducing Foursquare-Powered Place Pins on Pinterest." 2013. *Foursquare*, November 22. http://foursquare.tumblr.com/post/67672365878/introducing-foursquare-powered-place-pins-on-pinterest

Iosifidis, Petros. 2011. *Global Media and Communication Policy*. Houndmills, Basingstoke, Hampshire: Palgrave Macmillan.

Isaac, Mike. 2012. "With New Merchant Local Updates Tool, Foursquare Is Getting Serious about Its Business." *All Things D*, July 18. http://allthingsd.com/20120718/with-new-merchant-local-updates-tool-foursquare-is-getting-serious-about-its-business/

Isaac, Mike. 2013. "Foursquare's New App Is Open for Business." *All Things D*, January 29. http://allthingsd.com/20130129/foursquares-new-app-is-open-for-business/

ITU. 2013. *The World in 2013: ICT Facts and Figures*. Geneva: International Telecommunications Union (ITU).

ITU. 2016. *ICT Facts and Figures 2016*. Geneva: International Telecommunications Union (ITU).

ITU. 2017. *ICT Facts and Figures 2017*. Geneva: International Telecommunications Union (ITU).

Iwatani Kane, Yukari. 2010. "Apple's Quattro Deal Pressures Google." *Wall Street Journal*, January 6. http://www.wsj.com/articles/SB10001424052748703436504574640201846055072

Johns, Nicholas A. 2017. *The Age of Sharing*. Cambridge: Polity.

Johnson, Khari. 2018. "Foursquare Launches Places API for Startups and Small Businesses." *VentureBeat*, April 12. https://venturebeat.com/2018/04/12/foursquare-launches-places-api-for-startups-and-small-businesses/

Johnston, Anna. 2017. "Mobiles, Metadata and the Meaning of 'Personal' Information." *SalingerPrivacy*, January 19. https://www.salingerprivacy.com.au/2017/01/19/federalcourtdecision/

Jones, Kipp. 2015. "Skyhook under the Hood: Determining Location with Wi-Fi Access Points." *Skyhook Wireless*, March 11. http://blog.skyhookwireless.com/company/skyhook-under-the-hhod-determining-location-with-wi-fi-access-points

Kafer, Gary. 2016. "Documenting the Invisible: Political Agency in Trevor Paglen's Limit Telephotography." *Contemporaneity: Historical Presence in Visual Culture* 5 (1): 53–71.

Kapidzic, Sanja, and Nicole Martins. 2015. "Mirroring the Media: The Relationship Between Media Consumption, Media Internalization, and Profile Picture Characteristics on Facebook." *Journal of Broadcasting & Electronic Media* 59 (2): 278–297.

Kaplan, David. 2015. "Apple Maps Is No Longer a Punchline." *Geomarketing*, April 7. http://www.geomarketing.com/apple-maps-is-no-longer-a-punchline

Kaplan, David. 2018. "Tinder Test Place-Based Dating Feature with Foursquare, Mapbox." *GeoMarketing*, May 24. https://geomarketing.com/tinder-tests-place-based-dating-feature-with-foursquare-mapbox

Kardashian West, Kim (@KimKardashian). 2016. "Twitter Is Where I Can Freely Talk." Twitter, March 14, 2016, 12:26 p.m. https://twitter.com/KimKardashian/status/709460790219116545

Katz, James E, and Satomi Sugiyama. 2005. "Mobile Phones as Fashion Statements: The Co-creation of Mobile Communication's Public Meaning." In *Mobile Communications: Re-negotiation of the Social Sphere*, edited by Rich Ling and Per E. Pedersen, 63–81. London: Springer.

Kelly, Anne Marie. 2012. "Three Reasons Why Foursquare's New Advertising Model Might Work." *Forbes*, August 22. https://www.forbes.com/sites/annemariekelly/2012/08/22/three-reasons-why-foursquares-new-advertising-model-might-work/#5b29cf1121d8

Kennedy, Helen, Dag Elgesem, and Cristina Miguel. 2017. "On Fairness: User Perspectives on Social Media Data Mining." *Convergence: The International Journal of Research into New Media Technologies* 23 (3): 270–288.

Kennedy, Helen, Thomas Poell, and José van Dijck. 2015. "Data and Agency." *Big Data & Society* (July–December). https://doi.org/10.1177/2053951715621569

Kennedy, Jenny. 2016. "Conceptual Boundaries of Sharing." *Information, Communication & Society* 19 (4): 461–474.

Kerr, Dara. 2012. "Foursquare Launches Rating System, Competes with Yelp." *CNet*, November 5. https://www.cnet.com/news/foursquare-launches-rating-system-competes-with-yelp/

Khanna, Aran. 2015. "Stalking Your Friends with Facebook Messenger." *Medium*, May 26. https://medium.com/faith-and-future/stalking-your-friends-with-facebook-messenger-9da8820bd27d

Khazan, Olga. 2013. "The Creepy, Long-standing Practice of Undersea Cable Tapping." *The Atlantic*, July 16. https://www.theatlantic.com/international/archive/2013/07/the-creepy-long-standing-practice-of-undersea-cable-tapping/277855/

Kilday, Bill. 2018. *Never Lost Again: The Google Mapping Revolution That Sparked New Industries and Augmented Our Reality*. New York: HarperCollins.

Kim, Hyang-sook. 2016. "What Drives You to Check In on Facebook? Motivations, Privacy Concerns, and Mobile Phone Involvement for Location-Based Information Sharing." *Computers in Human Behavior* 54: 397–406.

Kim, Yoo-chul. 2012. "Will Google 'Koreanize' Its Privacy Policy?" *Korea Times*, March 29. http://www.koreatimes.co.kr/www/news/tech/2012/03/129_107970.html

King, Ritchie S., and Mika Gröndahl. 2012. "How Google Collected Data from Wi-Fi Networks." *The New York Times*, May 23. http://www.nytimes.com/interactive/2012/05/23/business/How-Google-Collected-Data-From-Wi-Fi-Networks.html

Kitchin, Rob, Claudio Colette, Leighton Evans, and Liam Heaphy. 2019. "Creating Smart Cities." In *Creating Smart Cities*, edited by Claudio Colette, Leighton Evans, Liam Heaphy, and Rob Kitchin, 1–18. London: Routledge.

Kitchin, Rob, Tracey P. Lauriault, and Matthew W. Wilson, eds. 2017. *Understanding Spatial Media*. London: SAGE.

Kloc, Joe. 2013. "Forget PRISM: FAIRVIEW Is the NSA's Project to "Own the Internet."" *The Daily Dot*, July 12. https://www.dailydot.com/news/fairview-prism-blarney-nsa-internet-spying-projects/

Kolodny, Lora. 2017. "Apple Is Suddenly Looking to Hire a Bunch of Map Tech Experts." *CBNC*, August 2. https://www.cbnc.com/2017/08/02/apple-has-over-70-map-tech-job-openings.html

Koops, Bert-Jaap, and Ronald Leenes. 2014. "Privacy Regulation Cannot Be Hardcoded: A Critical Comment on the 'Privacy by Design' Provision in Data-Protection Law." *International Review of Law, Computers & Technology* 28 (2): 159–171.

Krasnova, Hanna, Natasha F. Veltri, and Oliver Günther. 2012. "Self-disclosure and Privacy Calculus on Social Networking Sites: The Role of Culture." *Business & Information Systems Engineering* 3: 127–135.

Kuehn, Kathleen M. 2016. "Branding the Self on Yelp: Consumer Reviewing as Image Entrepreneurship." *Social Media + Society*. https://doi.org/10.1177/2056305116678895

Kurgan, Laura. 2013. *Close Up at a Distance: Mapping, Technology, and Politics*. Cambridge, MA: MIT Press.

Lacy, Sarah. 2009. *Once You're Lucky, Twice You're Good: The Rebirth of Silicon Valley and the Rise of Web 2.0*. New York: Gotham Books.

Lacy, Sarah. 2014. "Follow the Photos: The Real Reason Facebook Just Paid Almost 10% of Its Market Cap for WhatsApp." *PandoDaily*, February 19. http://pando.com/2014/02/19/follow-the-photos-the-real-reason-facebook-just-paid-almost-10-of-its-market-cap-for-whatsapp/

Lampinen, Ari, Vilma Lehtinen, Asko Lehmuskallio, and Sakari Tamminen. 2011. "We're in It Together: Interpersonal Management of Disclosure in Social Network Services." Paper presented at CHI 2011, Vancouver, BC, Canada, May 7–12.

Landau, Susan. 2016. "Choices: Privacy and Surveillance in a Once and Future Internet." *Dædalus: Journal of the American Academy of Arts & Sciences* 145 (1): 54–64.

Lapenta, Francesco. 2011. "Geomedia: On Location-Based Media, the Changing Status of Collective Image Production and the Emergence of Social Navigation Systems." *Visual Studies* 26 (1): 14–24.

Lardinois, Frederic. 2010. "Report: Location Sharing Is Coming to Facebook." *ReadWrite*, March 9. http://readwrite.com/2010/03/09/location_sharing_is_coming_to_facebook_-_how_will_users_react

Larkin, Brian. 2013. "The Politics and Poetics of Infrastructure." *The Annual Review of Anthropology* 42: 327–343.

Lavoie, Andrew. 2009. "The Online Zoom Lens: Why Internet Street-Level Mapping Technologies Demand Reconsideration of the Modern-Day Tort Notion of 'Public Privacy.'" *Georgia Law Review* 43, no. 2 (Winter): 575–616.

Lawrence, Amanda, Julian Thomas, John Houghton, and Paul Weldon. 2015. "Collecting the Evidence: Improving Access to Grey Literature and Data for Public Policy and Practice." *Australian Academic and Research Libraries* 46 (4): 229–249.

Lazzarato, Maurizio. 1996. "Immaterial Labor." In *Radical Thought in Italy: A Potential Politics*, edited by Paulo Virno and Michael Hardt, 142–157. Minneapolis: University of Minnesota Press.

Leaver, Tama, and Clare Lloyd. 2015. "Seeking Transparency in Locative Media." In *Locative Media*, edited by Rowan Wilken and Gerard Goggin, 162–174. New York: Routledge.

Lee, Newton. 2013. *Facebook Nation: Total Information Awareness*. New York: Springer.

Lehtiniemi, Tuukka, and Yki Kortesniemi. 2017. "Can the Obstacles to Privacy Self-management be Overcome? Exploring the Consent Intermediary Approach." *Big Data & Society* (July–December). https://doi.org/10.1177/2053951717721935

Lemos, André. 2009. "Locative Media in Brazil." *Wi: Journal of Mobile Media* 10. http://wi.hexagram.ca/?p=60

Leorke, Dale. 2015. "Location-Based Gaming Apps and the Commercialization of Locative Media." In *Mobility and Locative Media: Mobile Communication in Hybrid Spaces*, edited by Adriana de Souza e Silva and Mimi Sheller, 132–148. New York: Routledge.

Leorke, Dale. 2017. "'Know Your Place': Headmap Manifesto and the Vision of Locative Media." *Fibreculture Journal* 29. http://twentynine.fibreculturejournal.org/fcj-216-know-your-place-headmap-manifesto-and-the-vision-of-locative-media/

Leorke, Dale. 2019. *Location-Based Gaming: Play in Public Space*. London: Palgrave Macmillan.

Lessin, Jessica E. 2013. "Exclusive: Apple Buys (Another) Map App, Embark." *jessicalessin.com*, August 22. http://jessicalessin.com/2013/08/22/exclusive-apple-buys-another-map-app-embark/

Leszczynski, Agnieszka. 2012. "Situating the Geoweb in Political Economy." *Progress in Human Geography* 36 (1): 72–89.

Leszczynski, Agnieszka. 2015. "Spatial Big Data and Anxieties of Control." *Environment and Planning D: Society and Space* 33 (6): 965–984.

Leszczynski, Agnieszka. 2017. "Geoprivacy." In *Understanding Spatial Media*, edited by Rob Kitchin, Tracey P. Lauriault, and Matthew W. Wilson, 235–244. London: Sage.

Leszczynski, Agnieszka, and Jeremy Crampton. 2016. "Introduction: Spatial Big Data and Everyday Life." *Big Data & Society* (July–December): 1–6. https://doi.org/10.1177/2053951716661366

"Level Up: Your Foursquare Badges Now Show Your Expertise." 2011. *About Foursquare*, November 14. http://aboutfoursquare.com/level-up-your-foursquare-badges-now-show-your-expertise

Levine, Yasha. 2015. "The CIA Helped Sell a Mapping Startup to Google. Now They Won't Tell Us Why." *Pando*, July 1. https://pando.com/2015/07/01/cia-foia-google-keyhole/

Levinson, Alex. 2011. "3 Major Issues with the Latest iPhone Tracking 'Discovery.'" *Alex Levinson*, April 21. https://alexlevinson.wordpress.com/2011/04/21/3-major-issues-with-the-latest-iphone-tracking-discovery/

Levy, Steven. 2016. "Marsbot Is Dreaming of You." *Wired*, July 27. https://
www.wired.com/2016/07/marsbot-is-dreaming-of-you/

Lewis, Michael. 2001. *The New Thing*. New York: Penguin.

Lewis, Tania. 2008. *Smart Living: Lifestyle Media and Popular Expertise*.
New York: Peter Lang.

Leyden, John. 2014. "Angry Birds Developers Downplay Fresh Data Leak Claims."
The Register, April 1. https://www.theregister.co.uk/2014/04/01/angry_birds_
privacy_flap/

Li, Nan, and Guanling Chen. 2010. "Sharing Location in Online Social Networks."
IEEE Network 24 (4): 20–25. https://ieeexplore.ieee.org/document/
5578914/

Liaw, Adam (@adamliaw). 2018. "How I Organise My Messaging." Twitter, April 19,
2018, 3:21 p.m. https://twitter.com/adamliaw/status/987199471363616768

Licoppe, Christian. 2014. "Living Inside Location-Aware Mobile Social
Information: The Pragmatics of Foursquare Notifications." In *Living Inside
Mobile Social Information*, edited by James E. Katz, 109–130. Dayton,
OH: Greyden Press.

Licoppe, Christian, and Yoriko Inada. 2006. "Emergent Uses of a Multiplayer
Location-Aware Mobile Game: The Interactional Consequences of Mediated
Encounters." *Mobilities* 1 (1): 39–61.

Light, Ben. 2014. *Disconnecting with Social Networking Sites*. Houndmills, Basingstoke,
Hampshire: Palgrave Macmillan.

Lin, Jialiu, Shahriyar Amini, Jason I. Hong, Norman Sadeh, Janne Lindqvist, and Joy
Zhang. 2012. "Expectation and Purpose: Understanding Users' Mental Models
of Mobile App Privacy through Crowdsourcing." UbiComp '12, Proceedings
of the 2012 ACM Conference on Ubiquitous Computing, 501–510. https://
dl.acm.org/citation.cfm?id=2370290

Lindqvist, Janne, Justin Cranshaw, Jason Wiese, Jason Hong, and John Zimmerman.
2011. "I'm the Mayor of My House: Examining Why People Use Foursquare—
A Social-Driven Location Sharing Application." CHI 2011, Vancouver, BC,
Canada.

Ling, Rich, and Jonathan Donner. 2009. *Mobile Communication*. Cambridge: Polity.

Ling, Rich, and Leslie Haddon. 2003. "Mobile Telephony, Mobility and the
Coordination of Everyday Life." In *Machines That Become Us: The Social Context
of Personal Communication Technology*, edited by James E. Katz, 245–265. New
Brunswick, NJ: Transaction.

Ling, Rich, and Birgitte Yttri. 2002. "Hyper-coordination via Mobile Phones in
Norway." In *Perpetual Contact: Mobile Communication, Private Talk, Public
Performance*, edited by James E. Katz and Mark A. Aakhus, 139–169.
Cambridge: Cambridge University Press.

Lingis, Alphonso. 1994. *The Community of Those Who Have Nothing in Common*.
Minneapolis: University of Minnesota Press.

Litt, Eden, and Eszter Hargittai. 2016. "The Imagined Audience on Social Network
Sites." *Social Media + Society*. https://doi.org/10.1177/2056305116633482

Liu, Alan. 2004. "Transcendental Data: Toward a Cultural History and Aesthetics of
the New Encoded Discourse." *Critical Inquiry* 31 (1): 49–84.

Livingston, Jared S. 2011. "Invasion Contracts: The Privacy Implications of Terms
of Use Agreements in the Online Social Media Setting." *Albany Law Journal of
Science & Technology* 21 (3): 591–636.

Lobato, Ramon, and Julian Thomas. 2014. "Informal Media Economies." In *The Routledge Companion to Mobile Media*, edited by Gerard Goggin and Larissa Hjorth, 114–122. New York: Routledge.

Lobato, Ramon, and Julian Thomas. 2015. *The Informal Media Economy*. Cambridge: Polity.

LocationSmart. 2015. "LocationSmart and Locaid Announce Merger." *LocationSmart*, February 26. https://www.locationsmart.com/company/news/locationsmart-and-locaid-announce-merger

Lopez, Napier. 2015. "Foursquare's New Pinpoint Ad Platform Sells Ads Based on Users' Location Data." *The Next Web*, April 15. https://thenextweb.com/insider/2015/04/14/foursquares-new-pinpoint-ad-platform-sells-ads-based-on-users-location-data/

Low, Setha M., and Irwin Altman. 1992. "Place Attachment: A Conceptual Inquiry." In *Place Attachment*, edited by Irwin Altman and Setha M. Low, 1–12. New York: Plenum Press.

Lukermann, Fred. 1961. "The Concept of Location in Classical Geography." *Annals of the Association of American Geographers* 51: 194–210.

Lunden, Ingrid. 2011. "Skyhook CEO: 'Privacy Is the Third Rail of Location Services'; The WP7 Op." *Gigaom*, June 12. https://gigaom.com/2011/06/12/419-skyhook-ceo-privacy-is-the-third-rail-of-location-services-the-wp7-op/

Lupton, Deborah. 2013. *Risk*. 2nd ed. London: Routledge.

Lupton, Deborah. 2016. *The Quantified Self: A Sociology of Self-Tracking*. Cambridge: Polity.

Lynley, Matthew. 2016. "Pinterest Acquires the Team behind Highlight and Shorts." *TechCrunch*, July 14. https://techcrunch.com/2016/07/14/pinterest-acquires-the-team-behind-highlight-and-shorts/

Lyon, David. 2014. "Surveillance, Snowden, and Big Data: Capacities, Consequences, Critique." *Big Data & Society* (July–December): 1–13. https://doi.org/10.1177/2053951714541861

Maas, Willem. 2013. "Freedom of Movement Inside 'Fortress Europe.'" In *Global Surveillance and Policing: Borders, Security, Identity*, edited by Elia Zureik and Mark B. Salter, 233–245. London: Routledge.

MacAskill, Ewen, and Alex Hern. 2018. "Snowden Revelations Continue to Resonate." *The Guardian Weekly*, June 8: 4–5.

MacAskill, Ewen, and Dominic Rushe. 2013. "Snowden Document Reveals Key Role of Companies in NSA Data Collection." *The Guardian*, November 2. http://www.theguardian.com/world/2013/nov/01/nsa-data-collection-tech-firms

Macherey, Pierre. 2006. *A Theory of Literary Production*. Translated by Geoffrey Wall. London: Routledge.

Mackenzie, Adrian. 2010. *Wirelessness: Radical Empiricism in Network Cultures*. Cambridge, MA: MIT Press.

Mackenzie, Adrian. 2017. *Machine Learners: Archaeology of a Data Practice*. Cambridge, MA: MIT Press.

Macmillan, Duncan. 1995. "Introduction." In *Wood Notes Wild: Essays on the Poetry and Art of Ian Hamilton Finlay*, edited by Alec Finlay, 1–5. Edinburgh: Polygon.

Madrigal, Alexis C. 2012. "How Google Builds Its Maps—and What It Means for the Future of Everything." *The Atlantic*, September 6. https://www.theatlantic.com/technology/archive/2012/09/how-google-builds-its-maps-and-what-it-means-for-the-future-of-everything/261913/

Magistretti, Bérénice. 2017. "MomentFeed Raises $16 Million to Help Brands with Social Media Marketing." *VentureBeat*, February 23. https://venturebeat.com/2017/02/23/momentfeed-raises-16-3-million-to-help-brands-with-social-media-marketing/

Maldoff, Gabe, and Omer Tene. 2017. "Putting Data Benefits in Context: A Response to Kift and Nissenbaum." *I/S: A Journal of Law and Policy for the Information Society* 13 (2): 383–397.

Malik, Om. 2007. "Nokia Buys Navteq for $8 Billion, Bets Big on Location-Based Services." *Gigaom*, October 1. https://gigaom.com/2007/10/01/nokia-navteq/

Malik, Om. 2008. "How Placebase Survived Google Maps." *Gigaom*, May 21. https://gigaom.com/2008/05/21/placebase/

Malik, Om. 2012. "Here Is Why Facebook Bought Instagram." *Gigaom*, April 9. http://gigaom.com/2012/04/09/here-is-why-did-facebook-bought-instagram/

Mangla, Karan. 2012. "Under the Hood: Building the Location API." *Facebook Engineering*, March 9. https://www.facebook.com/notes/facebook-engineering/under-the-hood-building-the-location-api/10150558607303920

Mansell, Robin. 2004. "Political Economy, Power and New Media." *New Media & Society* 6 (1): 96–105.

Margulis, Stephen T. 2003. "On the Status and Contribution of Westin's and Altman's Theories of Privacy." *Journal of Social Issues* 59 (2): 411–429.

Margulis, Stephen T. 2011. "Three Theories of Privacy: An Overview." In *Privacy Online: Perspectives on Privacy and Self-Disclosure in the Social Web*, edited by Sabine Trepte and Leonard Reinecke, 9–17. Heidelberg: Springer-Verlag.

Mark Zuckerberg: Inside Facebook. 2013. Television program, September 22. Sydney: ABC 1.

Markowitz, Eric. 2012. "Meet 3 Start-ups behind Apple's New Maps." *Inc.*, June 13. http://www.inc.com/eric-markowitz/start-ups-behind-the-new-apple-maps.html

Marks, Paul. 2010. "Innovation: The Wi-Fi Database That Shamed Google." *New Scientist*, April 30. https://www.newscientist.com/article/dn18844-innovation-the-wi-fi-database-that-shamed-google/

Marshall, Matt. 2011. "Apple Buys C3 Technologies—A Shot at Google Maps." *VentureBeat*, October 29. http://venturebeat.com/2011/10/29/apple-buys-c3-technologies-one-more-cut-into-google-maps-yoke/

Martin, Fran. 2003. "Introduction to Section 4: Everyday Practices." In *Interpreting Everyday Culture*, edited by Fran Martin, 154–158. London: Hodder Arnold.

Marwick, Alice. 2013. *Status Update: Celebrity, Publicity, & Branding in the Social Media Age*. New Haven, CT: Yale University Press.

Marwick, Alice, and danah boyd. 2011. "'I Tweet Honestly, I Tweet Passionately': Twitter Users, Context Collapse, and the Imagined Audience." *New Media & Society* 13 (1): 114–133.

Marwick, Alice, and danah boyd. 2014. "Networked Privacy: How Teenagers Negotiate Context in Social Media." *New Media & Society* 16 (7): 1051–1067.

Marwick, Alice, and danah boyd. 2018. "Understanding Privacy at the Margins—Introduction." *International Journal of Communication* 12: 1157–1165.

Massey, Doreen. 2005. *For Space*. London: Routledge.

Mattern, Shannon. 2015. "Deep Time of Media Infrastructure." In *Signal Traffic: Critical Studies of Media Infrastructures*, edited by Lisa Parks and Nicole Starosielski, 94–112. Urbana: University of Illinois.

Mattern, Shannon. 2017. *Code and Clay . . . Data and Dirt: Five Thousand Years of Urban Media*. Minneapolis: University of Minnesota Press.

Mayer, Colin. 2004. "The Financing and Governance of New Technologies."
In *Financial Systems, Corporate Investment in Innovation, and Venture
Capital*, edited by Anthony Bartzokas and Sunil Mani, 32–51.
Cheltenham: Edward Elgar.

Mayer, Marissa. 2011. "Google Just Got Zagat Rated!" *Google Official Blog*, September
8. https://googleblog.blogspot.com/2011/09/google-just-got-zagat-rated.html

Mayer-Schönberger, Victor. 2009. *Delete: The Virtue of Forgetting in the Digital Age*.
Princeton, NJ: Princeton University Press.

Mayer-Schönberger, Victor, and Kenneth Cukier. 2013. *Big Data: A Revolution That
Will Transform How We Live, Work, and Think*. New York: Houghton Mifflin
Harcourt.

Mayfield, Kendra. 2002. "Route of Problem: Bad Online Maps." *Wired*, September 16.
https://www.wired.com/2002/09/route-of-problem-bad-online-maps/

Mazzucato, Mariana. 2013. *The Entrepreneurial State: Debunking Public vs. Private
Sector Myths*. London: Anthem Press.

McClendon, Brian. 2013. "Google Maps and Waze, Outsmarting Traffic Together."
Google Official Blog, June 11. https://googleblog.blogspot.com/2013/06/google-
maps-and-waze-outsmarting.html

McCosker, Anthony. 2015a. "Drone Media: Unruly Systems, Radical Empiricism and
Camera Consciousness." *Culture Machine* 16. http://www.culturemachine.net

McCosker, Anthony. 2015b. "Drone Vision, Zones of Protest, and the New Camera
Consciousness." *Media Fields Journal* 9.

McCosker, Anthony. 2017a. "Data Literacies for the Postdemographic Social Media
Self." *First Monday* 22 (10). http://firstmonday.org/ojs/index.php/fm/article/
view/7307/6550

McCosker, Anthony. 2017b. "Tagging Depression: Social Media and the Segmentation
of Mental Health." In *Digital Media: Transformations in Human Communication*,
edited by Paul Messaris and Lee Humphreys, 31–39. New York: Peter Lang.

McCosker, Anthony. 2017c. "Social Media Work: Reshaping Organisational
Communications, Extracting Digital Value." *Media International Australia*
163: 122–136.

McCosker, Anthony, and Rowan Wilken. 2017. "'Things That Should Be Short': Perec,
Sei Shōnagon, Twitter, and the Uses of Banality." In *The Afterlives of
Georges Perec*, edited by Rowan Wilken and Justin Clemens, 136–153.
Edinburgh: Edinburgh University Press.

McCullough, Malcolm. 2006. "On the Urbanism of Locative Media." *Places* 18 (2): 26–29.

McDermott, John. 2014. "Why the Web's Biggest Players Are Gobbling Up
Location-Based Apps." *Digiday*, February 20. http://digiday.com/platforms/
apple-google-microsoft-yahoo-are-betting-on-mobile/

McDermott, Yvonne. 2017. "Conceptualising the Right to Data Protection in an Era
of Big Data." *Big Data & Society* (January–June): 1–7. https://doi.org/10.1177/
2053951716686994

McDonald, Aleecia M., and Lorrie Faith Cranor. 2008. "The Cost of Reading
Privacy Policies." *I/S: A Journal of Law and Policy for the Information Society* 4
(3): 540–565.

McGowen, Jana. 2010. "Your Boring Life, Now Available Online: Analyzing
Google Street View and the Right to Privacy." *Texas Wesleyan Law Review* 3
(Spring): 477–494.

McQuire, Scott. 2016. *Geomedia: Networked Cities and the Future of Public Space*.
Cambridge: Polity.

McVeigh-Schultz, Joshua, and Nancy Baym. 2015. "Thinking of You: Vernacular Affordance in the Context of the Microsocial Relationship App, Couple." *Social Media + Society* (July–December): 1–13. https://doi.org/10.1177/2056305115604649

Meese, James, and Rowan Wilken. 2014. "Google Street View in Australia: Privacy Implications and Regulatory Solutions." *Media and Arts Law Review* 19 (4): 305–324.

Michaels, C. William. 2007. *No Greater Threat: America after September 11 and the Rise of a National Security State*. New York: Algora.

Microsoft. 2002. "Microsoft Completes Acquisition of Vicinity Corp." *Microsoft.com*, December 12. https://news.microsoft.com/2002/12/12/microsoft-completes-acquisition-of-vicinity-corp/

Microsoft. 2006. "Microsoft Acquires Vexcel Corp., a Worldwide Leader in Imagery and Remote Sensing Technology." *Microsoft.com*, May 4. https://news.microsoft.com/2006/05/04/microsoft-acquires-vexcel-corp-a-worldwide-leader-in-imagery-and-remote-sensing-technology/

Milgram, Stanley. 1992. "The Familiar Stranger: An Aspect of Urban Anonymity." In *The Individual in a Social World: Essays and Experiments*, edited by John Sabini and Maury Silver, 51–53. New York: McGraw-Hill.

Miller, Claire Cain. 2013. "Stern Words, and a Pea-Size Punishment, for Google." *New York Times*, April 22. https://www.nytimes.com/2013/04/23/business/global/stern-words-and-pea-size-punishment-for-google.html

Miller, Greg. 2014. "The Huge, Unseen Operation behind the Accuracy of Google Maps." *Wired*, December 8. https://www.wired.com/2014/12/google-maps-ground-truth/

Mims, Christopher. 2018. "Your Location Data Is Being Sold—Often Without Your Knowledge." *Wall Street Journal*, March 4. https://www.wsj.com/articles/your-location-data-is-being-soldoften-without-your-knowledge-1520168400

Mintz, Beth, and Michael Schwartz. 1985. *The Power Structure of American Business*. Chicago: University of Chicago Press.

Miranda, Maria. 2011. "Unsitely Aesthetics: The Reconfiguring of Public Space in Electronic Art." Panel presented at ISEA 2011, Istanbul. https://isea2011.sabanciuniv.edu/panel/unsitely-aesthetics

Miranda, Maria. 2013. *Unsitely Aesthetics*. Berlin: Errant Bodies Press.

Misa, Thomas. 2003. "The Compelling Tangle of Modernity and Technology." In *Modernity and Technology*, edited by Thomas Misa, Philip Brey, and Andrew Feenberg, 1–30. Cambridge, MA: MIT Press.

Mishra, Pankaj. 2014. "With $770k in Seed Funding, Playbasis Wants Businesses in Asia to Gamify." *TechCrunch*, February 16. http://techcrunch.com/2014/02/16/with-770k-in-seed-funding-playbasis-wants-businesses-in-asia-to-gamify/

Mlot, Stepanie. 2015. "New Systems Teach Driverless Cars to 'See.'" *PC Mag*, December 24. https://au.pcmag.com/software/40861/news/new-systems-teach-driverless-cars-to-see

Monmonier, Mark. 1996. *How to Lie with Maps*. 2nd ed. Chicago: University of Chicago Press.

Moore, Justin, Noah Weiss, Benjamin N. Lee, Max Elliot Sklar, and Blake Shaw. 2013. System and Method for Providing Recommendations with a Location-Based Service. US Patent 20130073422A1, filed March 8, 2012, and issued March 21, 2013.

Moore, W. G. ca. 1963. *A Dictionary of Geography*. Harmondsworth, Middlesex: Penguin.

Moores, Shaun. 2012. *Media, Place & Mobility*. Houndmills, Basingstoke, Hampshire: Palgrave Macmillan.

Morin, Dave. 2008. "Announcing Facebook Connect." *Facebook Developer Blog*, May 9. http://developers.facebook.com/blog/post/2008/05/09/announcing-facebook-connect/

Morley, David. 2000. *Home Territories: Media, Mobility and Identity*. London: Routledge.

Morley, David, and Kevin Robins. 1995. *Spaces of Identity: Global Media, Electronic Landscapes and Cultural Boundaries*. London: Routledge.

Mosco, Vincent. 2009. *The Political Economy of Communication*. 2nd ed. London: Sage.

Mosco, Vincent. 2014. *To the Cloud: Big Data in a Turbulent World*. Boulder, CO: Paradigm.

Mueller, Robert. 2017. "Google Adds New Tricks and Devices to Tango Augmented Reality." *Fast Company*, May 17. https://news.fastcompany.com/google-adds-new-tricks-and-devices-to-tango-augmented-reality-4038110

Munster, Anna. 2013. *An Aesthesia of Networks: Conjunctive Experience in Art and Technology*. Cambridge, MA: MIT Press.

Nagy, Peter, and Gina Neff. 2015. "Imagined Affordance: Reconstructing a Keyword for Communication." *Social Media + Society*. https://doi.org/10.1177/2056305115603385

Nardi, Bonnie A., and Vicki L. O'Day. 1999. *Information Ecologies: Using Technology with Heart*. Cambridge, MA: MIT Press.

Nath, Anjali. 2016. "Touched from Below: On Drones, Screens and Navigation." *Visual Anthropology* 29 (3): 315–330.

Neal, Andrew W. 2008. "Goodbye War on Terror? Foucault and Butler on Discourses of Law, War and Exceptionalism." In *Foucault on Politics, Security and War*, edited by Michael Dillon and Andrew W. Neal, 43–64. Houndmills, Basingstoke: Palgrave Macmillan.

Neff, Gina. 2012. *Venture Labor: Work and the Burden of Risk in Innovative Industries*. Cambridge, MA: MIT Press.

Neff, Gina, and Dawn Nafus. 2016. *Self-tracking*. Cambridge, MA: MIT Press.

Newsham, Jack. 2015a. "Skyhook and Google Settle a Long-Running Lawsuit." *Boston Globe*, March 9. https://www.bostonglobe.com/business/2015/03/09/skyhook-and-google-settle-long-running-lawsuit/jrWcEQ7Vn9AR4SBTnni1eI/story.html

Newsham, Jack. 2015b. "Skyhook Got $61 Million from Google Settlement." *Boston Globe*, May 19. https://www.bostonglobe.com/business/2015/05/19/skyhook-got-million-from-google-settlement/0q54ppSx2NqLrpyisw9ZZK/story.html

Newton, Casey. 2016a. "Foursquare Is Remaking Itself as a Bot." *The Verge*, May 24. https://www.theverge.com/2016/5/24/11763906/foursquare-marsbot-restaurant-recommendations-bots

Newton, Casey. 2016b. "Instagram Is Getting Rid of Photo Maps." *The Verge*, September 6. https://www.theverge.com/2016/9/6/12817340/instagram-photo-map-removals

Nippert-Eng, Christena. 1996. *Home and Work: Negotiating Boundaries through Everyday Life*. Chicago: University of Chicago Press.

Nippert-Eng, Christena. 2005. "Boundary Play." *Space & Culture* 8 (3): 302–324.

Nippert-Eng, Christena. 2010. *Islands of Privacy: Selective Concealment and Disclosure in Everyday Life*. Chicago: University of Chicago Press.

Nissenbaum, Helen. 2010. *Privacy in Context: Technology, Policy, and the Integrity of Social Life*. Stanford, CA: Stanford Law Books.

Nissenbaum, Helen. 2011. "A Contextual Approach to Privacy Online." *Dædalus: Journal of the American Academy of Arts & Sciences* 140 (4): 32–48.

Noble, Safiya Umoja. 2018. *Algorithms of Oppression: How Search Engines Reinforce Racism*. New York: New York University Press.

Norman, Donald A. 1990. *The Design of Everyday Things*. New York: Doubleday Business.

Norman, Donald A. 1998. *The Psychology of Everyday Things*. New York: Basic Books.

Nosowitz, Dan. 2010. "Apple Acquires Canadian 3D Mapping Software Company Poly9: Apple Maps?" *Fast Company*, July 14. http://www.fastcompany.com/1670332/apple-acquires-canadian-3d-mapping-software-company-poly9-apple-imaps

O'Beirne, Justin. 2017. "Google Maps's Moat: How Far Ahead of Apple Maps is Google Maps?" *Justinobeirne.com* (December). https://www.justinobeirne.com/google-maps-moat/#note9

O'Brien, Kevin J. 2010. "Google Data Admission Angers European Officials." *New York Times*, May 15. http://www.nytimes.com/2010/05/16/technology/16google.html?mcubz=0

O'Brien, Matt. 2015. "Uber Buys deCarta Mapping Service." *SiliconBeat*, March 3. http://www.siliconbeat.com/2015/03/03/uber-buys-decarta-mapping-service/

O'Dell, Jolie. 2010. "Facebook Location Features Confirmed and Coming Soon." *Mashable*, June 21. http://mashable.com/2010/06/21/facebook-location-confirmed/

O'Neill, Brian, and Michael Murphy. 2012. "Crossing Borders: The Introduction and Legislation of Satellite Radio in Canada." In *Down to Earth: Satellite Technologies, Industries, and Cultures*, edited by Lisa Parks and James Schwoch, 177–193. New Brunswick, NJ: Rutgers University Press.

O'Rourke, Karen. 2013. *Walking and Mapping: Artists as Cartographers*. Cambridge, MA: MIT Press.

OAIC. 2010. "Australian Privacy Commissioner Obtains Privacy Undertakings from Google." Google Street View Collection. Australian Government, Office of the Australian Information Commissioner, July 9. https://www.oaic.gov.au/media-and-speeches/statements/google-street-view-wi-fi-collection#australian-privacy-commissioner-obtains-privacy-undertakings-from-google

Obar, Jonathan A. 2015. "Big Data and *The Phantom Public*: Walter Lippmann and the Falacy of Data Privacy Self-management." *Big Data & Society* (July–December): 1–16. https://doi.org/10.1177/20539551715608876

Oliver, Julian. 2013. "No Network." *Julian Oliver*. https://julianoliver.com/no-network

Oliver, Julian. 2014a. "Border Bumping. In GSM We Trust." http://borderbumping.net/

Oliver, Julian. 2014b. "Stealth Infrastructure." *Rhizome.org*, May 20. http://rhizome.org/editorial/2014/may/20/stealth-infrastructure

Oliver, Julian (@julian0liver). 2016. "Art That Isn't Activism Decorates." Twitter, May 10, 2016, 9:41 a.m. https://twitter.com/julian0liver/status/729818632113500161

Oliver, Julian, Gordan Savičić, and Danja Vasiliev. 2011. "The Critical Engineering Manifesto." *The Critical Engineering Workshop Group*. https://criticalengineering.org/ce.pdf

Oliver, Sam. 2015. "How Coherent Navigation Can Help Apple with Location Technology and Talent." *AppleInsider*, May 18. http://appleinsider.com/articles/15/05/18/how-coherent-navigation-can-help-apple-with-location-technology-and-talent

OPC. 2011. "Google Inc. WiFi Data Collection." Office of the Privacy Commissioner of Canada / Commissariat à la protection de la vie privée du Canada. https://www.priv.gc.ca/en/opc-actions-and-decisions/investigations/investigations-into-businesses/2011/pipeda-2011-001/#findings

Orlowski, Andrew. 2012. "Google KNEW Street View Cars Were Slurping Wi-Fi: Wheels Fall Off 'One Rogue Engineer' Claim." *The Register*, April 30. https://www.theregister.co.uk/2012/04/30/google_slurp_ok/

Oxera Consulting. 2013. "What is the Economic Impact of Geo Services?" January. https://www.oxera.com/publications/what-is-the-economic-impact-of-geo-services/

Özkul, Didem. 2015a. "Location as a Sense of Place: Everyday Life, Mobile, and Spatial Practices in Urban Spaces." In *Mobility and Locative Media: Mobile Communication in Hybrid Spaces*, edited by Adriana de Souza e Silva and Mimi Sheller, 101–116. New York: Routledge.

Özkul, Didem. 2015b. "Mobile Communication Technologies and Spatial Perception: Mapping London." In *Locative Media*, edited by Rowan Wilken and Gerard Goggin, 39–51. New York: Routledge.

Özkul, Didem, and Lee Humphreys. 2015. "Record and Remember: Memory and Meaning-Making Practices through Mobile Media." *Mobile Media & Communication* 3 (3): 351–365.

Packer, Jeremy, and Stephen B. Crofts Wiley. 2012. "Introduction: The Materiality of Communication." In *Communication Matters: Materialist Approaches to Media, Mobility and Networks*, edited by Jeremy Packer and Stephen B. Crofts Wiley, 3–16. London: Routledge.

Packer, Jeremy, and Joshua Reeves. 2013. "Romancing the Drone: Military Desire and Anthropophobia from SAGE to Swarm." *Canadian Journal of Communication* 38: 309–331.

Paczkowski, John. 2013. "Apple Acquires Local Data Outfit Locationary." *All Things D*, July 19. http://allthingsd.com/20130719/apple-acquires-local-data-outfit-locationary/

Page, Xinru, and Alfred Kobsa. 2009. "Navigating the Social Terrain with Google Latitude." Proceedings of the 2010 iConference, Urbana-Champaign, Illinois, 174–178.

Paglen, Trevor. 2008. "Experimental Geography: From Cultural Production to the Production of Space." In *Experimental Geography: Radical Approaches to Landscape, Cartography, and Urbanism*, edited by Nato Thompson and Independent Curators International, 27–33. New York: Melville House.

Palmer, Maija. 2007. "TomTom Beats Garmin in Battle for Tele Atlas." *Financial Times*, November 17. https://www.ft.com/content/a0861026-9471-11dc-9aaf-0000779fd2ac

Panzarino, Matthew. 2013. "As Foursquare Concentrates on Demonstrating Value, It No Longer Allows Private Check-Ins on iOS7." *TechCrunch*, December 10. https://techcrunch.com/2013/12/09/as-foursquare-concentrates-on-demonstrating-value-it-no-longer-allows-private-check-ins-on-ios-7/

Panzarino, Matthew. 2014. "Foursquare Gets $15M and Licencing Deal from Microsoft to Power Location Context for Windows and Mobile." *TechCrunch* February 4. https://techcrunch.com/2014/02/04/foursquare-cuts-15m-deal-with-microsoft-to-power-location-and-context-for-windows-and-mobile/

Panzarino, Matthew. 2018. "Apple Is Rebuilding Maps from the Ground Up." *TechCrunch*, June 30. https://techcrunch.com/2018/06/29/apple-is-rebuilding-maps-from-the-ground-up/

Papacharissi, Zizi. 2010. "Conclusion: A Networked Self." In *A Networked Self: Identity, Community, and Culture on Social Network Sites*, edited by Zizi Papacharissi, 304–318. New York: Routledge.

Papalexakis, Evangelos, Konstantinos Pelechrinis, and Christos Faloutsos. 2014. "Spotting Misbehaviors in Location-based Social Networks using Tensors." WWW '14 Companion, Seoul, South Korea, April 7–11.

Pappas, Cleo, and Irene Williams. 2011. "Grey Literature: Its Emerging Importance." *Journal of Hospital Librarianship* 11 (3): 228–234.

Parikka, Jussi. 2010. *Insect Media: An Archaeology of Animals and Technology*. Minneapolis: University of Minnesota Press.

Parikka, Jussi. 2012. *What Is Media Archaeology?* Cambridge: Polity.

Parikka, Jussi. 2015. *A Geology of Media*. Minneapolis: University of Minnesota Press.

Park, Sora. 2017. *Digital Capital*. Houndmills, Basingstoke, Hampshire: Palgrave Macmillan.

Parks, Lisa. 2005. *Cultures in Orbit: Satellites and the Televisual*. Durham NC: Duke University Press.

Parks, Lisa. 2012. "Satellites, Oil, and Footprints: Eutelsat, Kazsat, and Post-Communist Territories in Central Asia." In *Down to Earth: Satellite Technologies, Industries, and Cultures*, edited by Lisa Parks and James Schwoch, 122–140. New Brunswick, NJ: Rutgers University Press.

Parks, Lisa. 2013. "Earth Observation and Signal Territories: Studying U.S. Broadcast Infrastructure through Historical Network Maps, Google Earth, and Fieldwork." *Canadian Journal of Communication* 38 (3): 285–308.

Parks, Lisa. 2014. "Drones, Infrared Imagery, and Body Heat." *International Journal of Communication* 8: 2518–2521. http://ijoc.org

Parks, Lisa. 2015. "Vertical Mediation: Geospatial Imagery and the US Wars in Afghanistan and Iraq." In *Mediated Geographies and Geographies of Media*, edited by Susan P. Mains, Julie Cupples, and Chris Lukinbeal, 159–175. Dordrecht: Springer.

Parks, Lisa. 2016. "Drones, Vertical Mediation, and the Targeted Class." *Feminist Studies* 42 (1): 227–235.

Parks, Lisa. 2017. "Vertical Mediation and the U.S. Drone War in the Horn of Africa." In *Life in the Age of Drone Warfare*, edited by Lisa Parks and Caren Kaplan, 134–157. Durham, NC: Duke University Press.

Parks, Lisa, and Caren Kaplan, eds. 2017. *Life in the Age of Drone Warfare*. Durham, NC: Duke University Press.

Parks, Lisa, and James Schwoch. 2012. "Introduction." In *Down to Earth: Satellite Technologies, Industries and Cultures*, edited by Lisa Parks and James Schwoch, 1–16. Piscataway, NJ: Rutgers University Press.

Parks, Lisa, and Nicole Starosielski. 2015a. "Introduction." In *Signal Traffic: Critical Studies of Media Infrastructures*, edited by Lisa Parks and Nicole Starosielski, 1–27. Urbana: University of Illinois Press.

Parks, Lisa, and Nicole Starosielski, eds. 2015b. *Signal Traffic: Critical Studies of Media Infrastructures*. Urbana: University of Illinois Press.

Parr, Ben. 2011. "Hootsuite Jumps into Location-Based Marketing with Acquisition of Geotoko." *Mashable*, October 12. http://mashable.com/2011/10/12/hootsuite-geotoko/#F_jztD7f8kqr

Pasquale, Frank. 2015. *The Black Box Society: The Secret Algorithms That Control Money and Information*. Cambridge, MA: Harvard University Press.

Patel, Nilay. 2011. "How Google Controls Android: Digging Deep into the Skyhook Filings." *The Verge*, May 12. https://www.theverge.com/2011/05/12/google-android-skyhook-lawsuit-motorola-samsung

Patil, Sameer, Apu Kapadia, Greg Norcie, and Adam J Lee. 2012. "'Check Out Where I Am!': Location-Sharing Motivations, Preferences, and Practices." CHI'12, Austin, Texas, May 5–10.

Patton, Paul. 1995. "Introduction." In Jean Baudrillard, *The Gulf War Did Not Take Place*. Translated by Paul Patton, 1–21. Sydney: Power Publications.

Paul, Christiane. 2013. "Contexts as Moving Targets: Locative Media Art and the Shifting Ground of Context Awareness." In *Throughout: Art and Culture Emerging with Ubiquitous Computing*, edited by Ulrik Ekman, 399–417. Cambridge, MA: MIT Press.

Perez, Sarah. 2014. "Apple Acquires Spotsetter, a Social Search Engine for Places." *TechCrunch*, June 6. http://techcrunch.com/2014/06/06/spotsetter-a-social-search-engine-for-places-acquired-by-apple/

Perez, Sarah. 2017. "Google Lens Will Let Smartphone Cameras Understand What They See and Take Action." *TechCrunch*, May 18. https://techcrunch.com/2017/05/17/google-lens-will-let-smartphone-cameras-understand-what-they-see-and-take-action/

Pertierra, Raul. 2012. "Disaporas, the New Media and the Globalized Homeland." In *Migration, Diaspora and Information Technology in Global Societies*, edited by Leopoldina Fortunati, Raul Pertierra, and Jane Vincent, 107–123. New York: Routledge.

Petronio, Sandra. 2002. *Boundaries of Privacy: Dialectics of Disclosure*. Albany: State University of New York Press.

Petronio, Sandra. 2007. "Translational Research Endeavors and the Practices of Communication Privacy Management." *Journal of Applied Communication Research* 35 (3): 218–222.

Pew Research Center. 2016. "Americans Increasingly Use Smartphones for More than Voice Calls, Texting." *PewResearchCenter*, January 29. http://www.pewresearch.org/fact-tank/2016/01/29/us-smartphone-use/ft_01-27-16_smartphoneactivities_640/

Pfanner, Eric. 2013. "Google Jousts with Wired South Korea over Quirky Internet Rules." *New York Times*, October 13. https://www.nytimes.com/2013/10/14/business/international/google-jousts-with-south-koreas-piecemeal-internet-rules.html

Phillips, David. 2011. "Identity and Surveillance Play in Hybrid Space." In *Online Territories: Globalization, Mediated Practice and Social Space*, edited by Miyase Christensen, André Jansson, and Christian Christensen, 171–184. New York: Peter Lang.

Pickett, S. T. A., and M. L. Cadenasso. 2002. "The Ecosystem as a Multidimensional Concept: Meaning, Model, and Metaphor." *Ecosystems* 5: 1–10.

Pinder, David. 2013. "Dis-locative Arts: Mobile Media and the Politics of Global Positioning." *Continuum: Journal of Media & Cultural Studies* 27 (4): 523–541.

Pink, Sarah, Debora Lanzeni, and Heather Horst. 2018. "Data Anxieties: Finding Trust in Everyday Digital Mess." *Big Data & Society* (January–June): 1–14. https://doi.org/10.1177/2053951718756685

Pink, Sarah, Shanti Sumartojo, Deborah Lupton, and Christine Heyes La Bond. 2017. "Mundane Data: The Routines, Contingencies and Accomplishments of Digital Living." *Big Data & Society*. https://doi.org/10.1177/2053951717700924

Placecast. 2017. "Transform Your Mobile Data into Mobile Customers." *Placecast*. http://placecast.net/index.html

Plantin, Jean-Christophe, Carl Lagoze, Paul N. Edwards, and Christian Sandvig. 2016. "Infrastructure Studies Meet Platform Studies in the Age of Google and Facebook." *New Media & Society* 20 (1): 293–310.

Pon, Bryan, Timo Seppälä, and Martin Kenney. 2014. "Android and the Demise of Operating System-Based Power: Firm Strategy and Platform Control in the Post-PC World." *Telecommunications Policy* 38: 979–991.

Pound, Roscoe. 1915. "Interests of Personality." *Harvard Law Review* 28 (4): 343–365.

Preuschat, Archibald. 2012. "TomTom Reaches Apple Maps Deal." *Wall Street Journal*, June 12. http://www.wsj.com/articles/SB10001424052702303901504577461822939015992

Prioleau, Marc. 2010. "Mobile Location Ecosystem [slideshare presentation]." *Prioleau Advisors*, November 17. https://www.slideshare.net/mprioleau/mobile-location-ecosystem

Prioleau, Marc. 2012a. "Google Maps Laying Down Landmines." *Prioleau Advisors*, June 6. http://prioleauadv.com/archives/301#more-301

Prioleau, Marc. 2012b. "New Kids on the Block: Apple Maps Launch." *Prioleau Advisors*, June 12. http://prioleauadv.com/archives/307#more-307

Prioleau, Marc. 2012c. "Let's Agree on What We Disagree About." *Prioleau Advisors*, September 24. http://prioleauadv.com/archives/351

Prioleau, Marc. 2013a. "Indoor Apple? Why Apple Bought WiFiSlam." *Prioleau Advisors*, March 24. http://prioleauadv.com/archives/443

Prioleau, Marc. 2013b. "Beating the Block: How to Take on Google Maps." *Prioleau Advisors*, June 16. http://prioleauadv.com/archives/473

PRNewswire. 2005. "Inrix Secures $6.1 Million in Oversubscribed Series A Round." *PRNewswire*, April 11. http://www.prnewswire.com/news-releases/inrix-secures-61-million-in-oversubscribed-series-a-round-54226452.html

Protalinski, Emil. 2011a. "Facebook Kills Places, but Emphasizes Location Sharing More." *ZDNet*, August 23. http://www.zdnet.com/blog/facebook/facebook-kills-places-but-emphasizes-location-sharing-more/2972

Protalinski, Emil. 2011b. "Gowalla Confirms: Facebook Acquires Location-based Social Network." *ZDNet*, December 5. http://www.zdnet.com/blog/facebook/gowalla-confirms-facebook-acquires-location-based-social-network/5808

Qiu, Jack Linchuan. 2016. *Goodbye iSlave: A Manifesto for Digital Abolition*. Urbana: University of Illinois Press.

Quain, John R. 2017. "What Self-driving Cars See." *New York Times*, May 25. https://www.nytimes.com/2017/05/25/automobiles/wheels/lidar-self-driving-cars.html

Rainie, Lee, and Barry Wellman. 2012. *Networked: The New Social Operating System*. Cambridge, MA: MIT Press.

Rakower, Lauren H. 2011. "Blurred Line: Zooming in on Google Street View and the Global Right to Privacy." *Brooklyn Journal of International Law* 37 (1): 317–347.

Rancière, Jacques. 2004a. *The Politics of Aesthetics: The Distribution of the Sensible*. Edited and translated by Gabriel Rockhill. London: Bloomsbury.

Rancière, Jacques. 2004b. "The Politics of Literature." *SubStance* 33 (1): 10–24.

Rancière, Jacques. 2009a. *Aesthetics and Its Discontents*. Translated by Steven Corcoran. Cambridge: Polity.

Rancière, Jacques. 2009b. *The Emancipated Spectator*. Translated by Gregory Elliott. London: Verso.

Rancière, Jacques. 2010. *Dissensus: On Politics and Aesthetics*. Edited and translated by Steven Corcoran. London: Continuum.

Rawlins, William K. 1992. *Friendship Matters: Communication, Dialectics, and the Life Course*. New Brunswick, NJ: Transaction Publishers.

Raynes-Goldie, Kate. 2010. "Aliases, Creeping and Wall Cleaning: Understanding Privacy in the Age of Facebook." *First Monday* 15 (1). http://firstmonday.org/ojs/index.php/fm/article/view/2775/2432

Reisinger, Don, and Shara Tibken. 2015. "Apple's Maps Gets Update, Complete with Transit Addition." *CNet*, June 9. http://www.cnet.com/au/news/apples-maps-gets-update-complete-with-transit-addition/

Relph, Ed. 1986. *Place and Placelessness*. Reprint and 3rd imprint. London: Pion.

Ren, Mai. 2011. "Location Cheating: A Security Challenge to Location-Based Social Network Services." Master of Science, Department of Computer Science and Engineering, University of Nebraska. http://digitalcommons.unl.edu/computerscidiss/31/

Restuccia, Francesco, Andrea Saracino, Sajal K. Das, and Fabio Martinelli. 2016. "LVS: A WiFi-Based System to Tackle Location Spoofing in Location-Based Services." IEEE 17th International Symposium on A World of Wireless, Mobile and Multimedia Networks (WoWMoM), Coimbra, Portugal, June 21–24.

Rhoen, Michiel. 2016. "Beyond Consent: Improving Data Protection through Consumer Protection Law." *Internet Policy Review: Journal of Internet Regulation* 5 (1): 1–15. https://doi.org/10.14763/2016.1.404

Richardson, Ingrid, and Rowan Wilken. 2009. "Haptic Vision, Footwork, Place-making: A Peripatetic Phenomenology of the Mobile Phone Pedestrian." *Second Nature: International Journal of Creative Media* 1 (2): 22–41.

Ritchie, Grant. 2012. "5 Big Map App Issues Apple Must Solve." *TechCrunch*, October 1. http://techcrunch.com/2012/09/30/5-problems-apple-needs-to-solve-in-its-maps-app/

Rockhill, Gabriel, and Philip Watts, eds. 2009. *History, Politics, Aesthetics: Jacques Rancière*. Durham, NC: Duke University Press.

Rosenblatt, Steven. 2017. "Unlocking the Power of Place for Marketers and Developers: Introducing Pilgrim SDK by Foursquare." *Medium*, March 1. https://medium.com/foursquare-direct/unlocking-the-power-of-place-for-marketers-and-developers-introducing-pilgrim-sdk-by-foursquare-ee879c502088

Rosol, Christoph. 2010. "From Radar to Reader: On the Origin of RFID." *Aether: The Journal of Media Geography* 5A: 37–49. http://geodata.csun.edu/~aether/pdf/volume_05a/rosol.pdf

Rossiter, Ned. 2015. "Locative Media as Logistical Media: Situating Infrastructure and the Governance of Labor in Supply-chain Capitalism." In *Locative Media*, edited by Rowan Wilken and Gerard Goggin, 208–223. New York: Routledge.

Rossiter, Ned. 2016. *Software, Infrastructure, Labor: A Media Theory of Logistical Nightmares*. London: Routledge.

Rossiter, Ned. 2017. "Imperial Infrastructures and Asia beyond Asia: Data Centres, State Formation and the Territorialities of Logistical Media." *Fibreculture Journal* 29. http://twentynine.fibreculturejournal.org/

Rubinstein, Ira S., and Nathaniel Good. 2013. "Privacy by Design: A Counterfactual Analysis of Google and Facebook Privacy Incidents." *Berkeley Technology Law Journal* 28: 1333–1413.

RT. 2013. "Germany Fines Google €145,000 over Personal Data Collection." *RT.com*, April 22. https://www.rt.com/news/personal-privacy-google-data-226/

Rueb, Teri. 2015. "Restless: Locative Media as Generative Displacement." In *Mobility and Locative Media: Mobile Communication in Hybrid Spaces*, edited by Adriana de Souza e Silva and Mimi Sheller, 241–258. New York: Routledge.

Ruppert, Evelyn, Engin Isin, and Didier Bigo. 2017. "Data Politics." *Big Data & Society* (July-December): 1–7. https://doi.org/10.1177/2053951717717749

Russo, Alexander, and Bill Kirkpatrick. 2012. "Beyond the Terrestrial?: Networked Distribution, Multimodal Media, and the Place of the Local in Satellite Radio." In *Down to Earth: Satellite Technologies, Industries, and Cultures*, edited by Lisa Parks and James Schwoch, 156–176. New Brunswick, NJ: Rutgers University Press.

Saker, Michael. 2017. "Foursquare and Identity: Checking-in and Presenting the Self through Location." *New Media & Society* 19 (6): 934–949.

Saker, Michael, and Leighton Evans. 2016. "Locative Mobile Media and Time: Foursquare and Technological Memory." *First Monday* 21, no. 2 (February). https://doi.org/10.5210/fm.v21i2.6006

Sakr, Naomi. 2012. "From Satellite to Screen: How Arab TV Is Shaped in Space." In *Down to Earth: Satellite Technologies, Industries, and Cultures*, edited by Lisa Parks and James Schwoch, 143–155. New Brunswick, NJ: Rutgers University Press.

Salmond, Michael. 2010. "The Power of Momentary Communities: Locative Media and (In)Formal Protest." *Aether: The Journal of Media Geography* 5A: 90–100. http://geogdata.csun.edu/~aether/pdf/volume_05a/salmond.pdf

Salter, Michael. 2016. "Privates in the Online Public: Sex(ting) and Reputation on Social Media." *New Media & Society* 18 (11): 2723–2739.

Sande, Steven. 2012. "Apple's Vector Maps Save Memory, Go Further When You're Offline." *Engadget*, October 5. https://www.engadget.com/2012/10/05/apples-vector-maps-go-further-offline/

Savage, Charlie. 2013. "In Test Project, NSA Tracked Cellphone Locations." *New York Times*, October 2. http://www.nytimes.com/2013/10/03/us/nsa-experiment-traced-us-cellphone-locations.html?mcubz=0

Savage, Charlie. 2016. "Obama Administration Set to Expand Sharing of Data That NSA Intercepts." *New York Times*, February 25. https://www.nytimes.com/2016/02/26/us/politics/obama-administration-set-to-expand-sharing-of-data-that-nsa-intercepts.html

Sawers, Paul. 2015. "Why 3 Car Giants Just Bought Nokia's Mapping Business for $3B." *VentureBeat*, August 3. https://venturebeat.com/2015/08/03/why-3-car-giants-just-bought-nokias-mapping-division-for-3b/

Scahill, Jeremy, and Josh Begley. 2015. "The Great SIM Heist: How Spies Stole the Keys to the Encryption Castle." *The Intercept*, February 20. https://theintercept.com/2015/02/19/great-sim-heist/

Scahill, Jeremey, and Glenn Greenwald. 2014. "The NSA's Secret Role in the U.S. Assassination Program." *The Intercept*, February 10. https://theintercept.com/2014/02/10/the-nsas-secret-role/

Schiller, Dan. 1999. "Deep Impact: The Web and the Changing Media Economy." *Info* 1, no. 1 (February): 35–51.

Schleser, Max, and Marsha Berry, eds. 2018. *Mobile Storytelling in an Age of Smartphones*. New York: Palgrave Macmillan.

Schonfeld, Erick. 2007. "Microsoft Buys Multimap for a Reported $50 Million." *Techcrunch*, December 13. https://techcrunch.com/2007/12/12/microsoft-buys-multimap-for-a-reported-50-million/

Schrock, Andrew R. 2015. "Communicative Affordances of Mobile Media: Portability, Availability, Locatability, and Multimediality." *International Journal of Communication* 9: 1229–1246.

Schutzberg, Adena. 2005. "The Technology behind Google Maps." *Directions Magazine*, February 20. http://www.directionsmag.com/entry/the-technology-behind-google-maps/123540

Schutzberg, Adena. 2006. "Microsoft and GeoTango." *Directions Magazine*, January 17. https://www.directionsmag.com/article/3035

Schutzberg, Adena, and Hal Reid. 2006. "Microsoft Addresses Vexcel Acquisition." *Directions Magazine*, May 10. https://www.directionsmag.com/article/2942

Schwartz, Raz. 2013. "The Networked Familiar Stranger: An Aspect of Online and Offline Urban Anonymity." In *Mobile Media Practices, Presence and Politics: The Challenge of Being Seamlessly Mobile*, edited by Kathleen M. Cumiskey and Larissa Hjorth, 135–149. New York: Routledge.

Schwartz, Raz. 2015. "Online Place Attachment: Exploring Technological Ties to Physical Places." In *Mobility and Locative Media: Mobile Communication in Hybrid Spaces*, edited by Adriana de Souza e Silva and Mimi Sheller, 85–100. New York: Routledge.

Schwartz, Raz, and Germaine R. Haleqoua. 2015. "The Spatial Self: Location-Based Identity Performance on Social Media." *New Media & Society* 17 (10): 1643–1660.

Scobie, Stephen. 1995. "Models of Order." In *Wood Notes Wild: Essays on the Poetry and Art of Ian Hamilton Finlay*, edited by Alec Finlay, 177–205. Edinburgh: Polygon.

Segall, Jordan E. 2010. "Google Street View: Walking the Line of Privacy-intrusion upon Seclusion and Publicity Given to Private Facts in the Digital Age." *Pittsburgh Journal of Technology Law & Policy* 10 (Spring). https://tlp.law.pitt.edu/ojs/index.php/tlp/article/view/51

Senft, Terri. 2008. *Camgirls: Celebrity and Community in the Age of Social Networks*. New York: Peter Lang.

Serrano, Alphonso. 2014. "Report: Spy Agencies Comb through Leaky Phone Apps for Personal Data." *Al Jazeera*, January 27. http://america.aljazeera.com/articles/2014/1/27/report-spy-agenciescombthroughphoneappsforpersonaldata.html

Shapiro, Carl, and Hal R. Varian. 1999. *Information Rules: A Strategic Guide to the Network Economy*. Boston, MA: Harvard Business Review Press.

Shaw, Blake. 2012. "Machine Learning with Large Networks of People and Places." *Vimeo* [video file]. http://vimeo.com/39088490

Shaw, Blake, Jon Shea, Siddhartha Sinha, and Andrew Hogue. 2012. "Learning to Rank for Spatiotemporal Search." Proceedings of the sixth ACM International Conference on Web Search and Data Mining (WSDM '13), Rome, Italy, February 4–8, 2012. http://www.metablake.com/foursquare/wsdm2013-final.pdf

Sheehan, John. 2013. "APIs Are Dead, Long Live APIs." *The Next Web*, March 13. https://thenextweb.com/dd/2013/03/12/apis-are-dead-long-live-apis/

Sheller, Mimi. 2017. "From Spatial Turn to Mobilities Turn." *Current Sociology* 65 (4): 623–639.

Sheridan, Joanna, and Kerry Chamberlain. 2011. "The Power of Things." *Qualitative Research in Psychology* 8 (4): 315–332.

Shklovski, Irina, Scott D. Mainwaring, Halla Hrund Skúladóttir, and Höskuldur Borgthorsson. 2014. "Leakiness and Creepiness in App Space: Perceptions of Privacy and Mobile App Use." CHI 2014, April 26–May 1, 2014, Toronto, ON.

Shontell, Alyson. 2011. "Foursquare Passes 1,000,000,000 Check-ins." *Business Insider*, September 20. http://articles.businessinsider.com/2011-09-20/tech/30179114_1_foursquare-check-ins-app

Showers, Carolin J. 2002. "Integration and Compartmentalization: A Model of Self-structure and Self-change." In *Advances in Personality Science*, edited by Daniel Cervone and Walter Mischel, 271–291. New York: Guilford.

Showers, Carolin J., and Virgil Zeigler-Hill. 2007. "Compartmentalization and Integration: The Evaluative Organization of Contextualized Selves." *Journal of Personality* 75 (6): 1181–1204.

Siegler, M. G. 2010. "Come on Yelp, Really? Dukes, Barons, and Kings of Venues?" *TechCrunch*, June 9. http://techcrunch.com/2010/06/09/yelp-royalty/

Silverstone, Roger. 2005. "The Sociology of Mediation and Communication." In *The Sage Handbook of Sociology*, edited by Craig Calhoun, Chris Rojek, and Bryan Turner, 188–207. London: Sage.

Simpson, John A., and Edmund S. C. Weiner. 1989. *The Oxford English Dictionary*, 2nd ed., Vol. VIII. Oxford: Clarendon Press.

Sinclair, John. 2012. *Advertising, The Media and Globalisation: A World in Motion*. London: Routledge.

Singh, Ishveena. 2018. "Insane, Shocking, Outrageous: Developers React to Changes in Google Maps API." *Geoawesomeness*, May 3. http://geoawesomeness.com/developers-up-in-arms-over-google-maps-api-insane-price-hike/

Sklar, Max, Blake Shaw, and Andrew Hogue. 2012. "Recommending Interesting Events in Real-time with Foursquare Check-ins." RecSys '12, Proceedings of the Sixth ACM Conference on Recommender Systems, Dublin, Ireland, September 9–13.

Skovholt, Karianne, and Jan Svennevig. 2006. "Email Copies in Workplace Interaction." *Computer-Mediated Communication* 12 (1): 42–65. https://doi.org/10.1111/j.1083-6101.2006.00314.x

Skyhook. 2014. "Skyhook Wireless Acquired by TruePosition." *Skyhook*, February 19. http://blog.skyhookwireless.com/company/trueposition-acquires-skyhook-wireless

SkyhookWireless. 2017. "Skyhook Privacy Policy." *SkyhookWireless*, January 25. http://www.skyhookwireless.com/privacy

Slater, Don. 1997. *Consumer Culture and Modernity*. Cambridge: Polity.

Slivka, Eric. 2015. "Apple Appears to Have Acquired GPS Firm Coherent Navigation [Confirmed]." *MacRumors*, May 17. http://www.macrumors.com/2015/05/17/apple-coherent-navigation-acquisition/

Smith, David C. 2011. Device for and Method of Geolocation. US Patent 7,893,875 B1, filed March 11, 2009, and issued February 22, 2011.

Smith, Harrison. 2017. "Metrics, Locations, and Lift: Mobile Location Analytics and the Production of Second-Order Geodemographics." *Information, Communication & Society*. https://doi.org/10.1080/1369118X.2017.1397726

Snow, Christopher, Darren Hayes, and Catherine Dwyer. 2016. "Leakage of Geolocation Data by Mobile Ad Networks." *Journal of Information Systems Applied Research* 9 (2): 24–33.

Solnit, Rebecca. 2014. *Encyclopedia of Trouble and Spaciousness*. San Antonio, TX: Trinity University Press.

Solove, Daniel J. 2004. *The Digital Person: Technology and Privacy in the Information Age*. New York: New York University Press.

Solove, Daniel J. 2006. "A Taxonomy of Privacy." *University of Pennsylvania Law Review* 154, no. 3 (January): 477–560.

Solove, Daniel J. 2007. *The Future of Reputation: Gossip, Rumor, and Privacy on the Internet*. New Haven, CT: Yale University Press.

Solove, Daniel J. 2008. *Understanding Privacy*. Cambridge, MA: Harvard University Press.

Solove, Daniel J. 2013. "Introduction: Privacy Self-management and the Consent Dilemma." *Harvard Law Review* 126 (7): 1880–1903.

Soltani, Ashkan, and Barton Gellman. 2013. "New Documents Show How the NSA Infers Relationships Based on Mobile Location Data." *Washington Post*, December 10. https://www.washingtonpost.com/news/the-switch/wp/2013/12/10/new-documents-show-how-the-nsa-infers-relationships-based-on-mobile-location-data/?utm_term=.5436d4eeda6e

Soltani, Ashkan, Andrea Peterson, and Barton Gellman. 2013. "NSA Uses Google Cookies to Pinpoint Targets for Hacking." *Washington Post*, December 10. https://www.washingtonpost.com/news/the-switch/wp/2013/12/10/nsa-uses-google-cookies-to-pinpoint-targets-for-hacking/?utm_term=.24a1eb06bdc9

Song, Ji-hye. 2011. "Google Amassed E-mails, Chats: Police." *Korea JoongAng Daily*, January 7. http://koreajoongangdaily.joins.com/news/article/article.aspx?aid=2930655

Sottek, T. C. 2012. "Facebook 'Find Friends Nearby' Feature Available for Web and Mobile." *The Verge*, June 24. https://www.theverge.com/2012/6/24/3114544/facebook-find-friends-nearby

Spiegel. 2013. "Privacy Scandal: NSA Can Spy on Smart Phone Data." *Spiegel Online*, September 7. http://www.spiegel.de/international/world/privacy-scandal-nsa-can-spy-on-smart-phone-data-a-920971-druck.html

Star, Susan Leigh. 1999. "The Ethnography of Infrastructure." *American Behavioral Scientist* 43 (3): 377–391.

Star, Susan Leigh, and Karen Ruhleder. 1996. "Steps Toward an Ecology of Infrastructure: Design and Access for Large Information Spaces." *Information Systems Research* 7 (1): 111–134.

Starosielski, Nicole. 2015a. "Fixed Flow: Undersea Cables as Media Infrastructure." In *Signal Traffic: Critical Studies of Media Infrastructures*, edited by Lisa Parks and Nicole Starosielski, 53–70. Urbana: University of Illinois Press.

Starosielski, Nicole. 2015b. *The Undersea Network*. Durham, NC: Duke University Press.

Sterling, Greg. 2010. "Google Ends Street View WiFi Data Collection, May Now Need Other Sources for Location." *Search Engine Land*, October 20. https://searchengineland.com/google-ends-street-view-wifi-data-collection-potentially-needs-other-sources-for-location-53373

Sterling, Greg. 2013a. "Facebook Nearby Is Now Facebook 'Local Search.'" *Search Engine Land*, April 3. https://searchengineland.com/facebook-nearby-is-now-facebook-local-search-154507

Sterling, Greg. 2013b. "Foursquare Adds Menu Items to Search Capability." *Search Engine Land*, September 9. https://searchengineland.com/foursquare-adds-menu-items-to-search-capability-171505

Sterling, Greg. 2015. "Mobile Marketer UberMedia Introduces the Concept of 'LROI.'" *Marketing Land*, December 2. https://marketingland.com/mobile-marketer-ubermedia-introduces-the-concept-of-lroi-154383

Sterling, Greg. 2016. "Foursquare Introduces 'Attribution' to Measure Offline Results." *Marketing Land*, February 23. https://marketingland.com/foursquare-introduces-attribution-product-to-measure-offline-results-165685

Stewart, Kathleen. 2007. *Ordinary Affects*. Durham, NC: Duke University Press.

Strachan, Lindsey. 2011. "Re-Mapping Privacy Law: How the Google Maps Scandal Requires Tort Law Reform." *Richmond Journal of Law and Technology* XVII (4): http://jolt.richmond.edu/v17i4/article14.pdf

Stroz Friedberg. 2010. "Source Code Analysis of gstumbler." *Stroz Friedberg*, June 3. https://static.googleusercontent.com/media/www.google.com/en//googleblogs/pdfs/friedberg_sourcecode_analysis_060910.pdf

Sturdevant, Rick W. 2012. "The NAVSTAR Global Positioning System: From Military Tool to Global Utility." In *Down to Earth: Satellite Technologies, Industries, and Cultures*, edited by Lisa Parks and James Schwoch, 99–121. New Brunswick, NJ: Rutgers University Press.

Su, Jimmy, Jinjian Zhai, and Tao Wei. 2014. "A Little Bird Told Me: Personal Information Sharing in Angry Birds and Its Ad Libraries." *FireEye*, March 27. https://www.fireeye.com/blog/threat-research/2014/03/a-little-bird-told-me-personal-information-sharing-in-angry-birds-and-its-ad-libraries.html

Summerhayes, Catherine. 2015. *Google Earth: Outreach and Activism*. London: Bloomsbury.

Sutko, Daniel M., and Adriana de Souza e Silva. 2011. "Location-Aware Mobile Media and Urban Sociability." *New Media & Society* 13 (5): 807–823.

Swan, Melanie. 2012. "Sensor Mania! The Internet of Things, Wearable Computing, Objective Metrics, and the Quantified Self 2.0." *Journal of Sensor and Actuator Networks* 1: 217–253.

Swisher, Kara. 2010. "Exclusive: Apple to Buy Quattro Wireless for $275 Million." *All Things D*, January 4. http://allthingsd.com/20100104/exclusive-apple-to-buy-quattro-wireless-for-275-million/

Szoldra, Paul. 2014. "Snowden: Here's Everything We've Learned in One Year of Unprecedented Top-Secret Leaks." *BusinessInsider*, June 7. https://www.businessinsider.com.au/snowden-leaks-timeline-2014-6?r=US&IR=T

Tacchi, Jo. 2006. "Studying Communicative Ecologies: An Ethnographic Approach to Information and Communication Technologies (ICTs)." Proceedings of the 56th Annual Conference of the International Communication Association, Dresden, Germany. http://eprints.qut.edu.au/4400/1/4400_1.pdf

Taddicken, Monika, and Cornelia Jers. 2011. "The Uses of Privacy Online: Trading a Loss of Privacy for Social Web Gratifications?" In *Privacy Online: Perspectives on Privacy and Self-Disclosure in the Social Web*, edited by Sabine Trepte and Leonard Reinecke, 225–255. New York: Plenum Press.

Taffel, Sy. 2013. "Scalar Entanglement in Digital Media Ecologies." *NECSUS: European Journal of Media Studies* (Spring). http://www.necsus-ejms.org/scalar-entanglement-in-digital-media-ecologies/

Tarkka, Minna. 2010. "Labours of Location: Acting in the Pervasive Media Space." In *The Wireless Spectrum: The Politics, Practices, and Poetics of Mobile Media*, edited by Barbara Crow, Michael Longford, and Kim Sawchuk, 131–145. Toronto: University of Toronto Press.

Tate, Ryan. 2014. "With Foursquare Deal, Microsoft Aims for Supremacy in Hyper-local Search." *Wired*, February 5. https://www.wired.com/2014/02/tracking-war-foursquare-microsoft/

Tawil-Souri, Helga. 2015. "Cellular Borders: Dis/Connecting Phone Calls in Israel-Palestine." In *Signal Traffic: Critical Studies of Media Infrastructures*, edited by Lisa Parks and Nicole Starosielski, 157–180. Urbana: University of Illinois Press.

Terranova, Tiziana. 2000. "Free Labor: Producing Culture for the Digital Economy." *Social Text* 18 (2): 33–58.

Thatcher, Jim. 2014. "Living on Fumes: Digital Footprints, Data Fumes, and the Limitations of Spatial Big Data." *International Journal of Communication* 8: 1765–1783.

Thielmann, Tristan. 2010. "Locative Media and Mediated Localities: An Introduction to Media Geography." *Aether: The Journal of Media Geography* 5A: 1–17.

Thrun, Sebastian, and John J. Leonard. 2008. "Simultaneous Localization and Mapping." In *Springer Handbook of Robotics*, edited by Bruno Siciliano and Oussama Khatib, 871–889. Berlin: Springer-Verlag.

Thussu, Daya Kishan. 2009. *Internationalizing Media Studies*. London: Routledge.

Timeto, Federica. 2015. "Locative Media, Performing Spatiality: A Nonrepresentational Approach to Locative Media." In *Locative Media*, edited by Rowan Wilken and Gerard Goggin, 94–106. New York: Routledge.

Torre, Paul. 2012. "Content vs. Delivery: The Global Battle for German Satellite Television." In *Down to Earth: Satellite Technologies, Industries, and Cultures*, edited by Lisa Parks and James Schwoch, 204–218. New Brunswick, NJ: Rutgers University Press.

Townsend, Anthony. 2006. "Locative-Media Artists in the Contested-Aware City." *Leonardo* 39 (4): 345–347.

Townsend, Anthony. 2013. *Smart Cities: Big Data, Civic Hackers, and the Quest for a New Utopia*. New York: W. W. Norton.

Trepte, Sabine, Leonard Reinecke, Nicole B. Ellison, Oliver Quiring, Mike Z. Yao, and Marc Ziegele. 2017. "A Cross-Cultural Perspective on the Privacy Calculus." *Social Media + Society* (January–March): 1–13.

Trottier, Daniel. 2018. "Privacy and Surveillance." In *The SAGE Handbook of Social Media*, edited by Jean Burgess, Alice Marwick, and Thomas Poell, 463–478. London: Sage.

Tseng, Erick. 2010. "Making Mobile More Social." *The Facebook Blog*, November 4. http://blog.facebook.com/blog.php?post=446167297130

Tsotsis, Alexia. 2012a. "Right before Acquisition, Instagram Closed $50m at a $500m Valuation from Sequoia, Thrive, Greylock and Benchmark." *TechCrunch*, April 9. http://techcrunch.com/2012/04/09/right-before-acquisition-instagram-closed-50m-at-a-500m-valuation-from-sequoia-thrive-greylock-and-benchmark/

Tsotsis, Alexia. 2012b. "Facebook Buys Location-Based Discovery App Glancee." *TechCrunch*, May 4. http://techcrunch.com/2012/05/04/facebook-buys-location-based-discovery-app-glancee/

Tsotsis, Alexia. 2012c. "Instagram 3.0 Bets Big on Geolocation with Photo Maps, Letting You Showcase the Story behind Your Photos." *TechCrunch*, August 16. http://techcrunch.com/2012/08/16/instagram-3-0-bets-big-on-geolocation-with-photo-maps-letting-you-showcase-the-story-behind-your-photos/

Tulloch, John, and Deborah Lupton. 2003. *Risk and Everyday Life*. London: Sage.

Tuters, Marc. 2012. "From Mannerist Situationism to Situated Media." *Convergence: The International Journal of Research into New Media Technologies* 18 (3): 267–282.

Tweney, Dylan. 2013. "South Korea's Heavily Regulated Internet Gives Google Conniptions." *VentureBeat*, October 14. https://venturebeat.com/2013/10/14/korea-internet-regulation/

Urry, John. 2007. *Mobilities*. Cambridge: Polity.

Vainikka, Eliisa, Elina Noppari, and Janne Seppänen. 2017. "Exploring Tactics of Public Intimacy on Instagram." *Participations: Journal of Audience & Reception Studies* 14 (1): 108–128. http://www.participations.org/Volume%2014/Issue%201/7.pdf

Van Alstyne, Marshall W., Geoffrey G. Parker, and Sangeet Paul Choudary. 2016. "Pipelines, Platforms, and the New Rules of Strategy." *Harvard Business Review* (April). https://hbr.org/2016/04/pipelines-platforms-and-the-new-rules-of-strategy

Van Couvering, Elizabeth. 2011. "Navigational Media: The Political Economy of Online Traffic." In *The Political Economics of Media: The Transformation of the Global Media Industries*, edited by Dwayne Winseck and Dal Yong Jin, 183–200. London: Bloomsbury.

van den Akker, Robin. 2015. "Walking in the Hybrid City: From Micro-coordination to Chance Orchestration." In *Mobility and Locative Media: Mobile Communication in Hybrid Spaces*, edited by Adriana de Souza e Silva and Mimi Sheller, 33–47. New York: Routledge.

van der Nagel, Emily. 2017. "Social Media Pseudonymity: Affordances, Practices, Disruptions." Doctor of Philosophy dissertation, Department of Media and Communication, Swinburne University of Technology.

van der Nagel, Emily. 2018. "Alts and Automediality: Compartmentalising the Self through Multiple Social Media Platforms." *M/C Journal* 21 (2). http://journal.media-culture.org.au/index.php/mcjournal/article/view/1379

van der Sloot, Bart, and Frederik Zuiderveen Borgesius. 2012. "Google and Data Protection." In *Google and the Law*, edited by Aurelio Lopez-Tarruella, 75–111. The Hague: T. M. C. Asser Press (Springer).

van Dijck, José. 2011. "Tracing Twitter: The Rise of a Microblogging Platform." *International Journal of Media and Cultural Politics* 7 (3): 333–348.

van Dijck, José. 2013a. *The Culture of Connectivity: A Critical History of Social Media*. Oxford: Oxford University Press.

van Dijck, José. 2013b. "Facebook and the Engineering of Connectivity: A Multilayered Approach to Social Media." *Convergence: The International Journal of Research into New Media Technologies* 19 (2): 141–155.

van Dijck, José. 2013c. "'You Have One Identity': Performing the Self on Facebook and LinkedIn." *Media Culture & Society* 35 (2): 199–215.

van Dijck, José, and Thomas Poell. 2013. "Understanding Social Media Logic." *Media and Communication* 1 (1): 2–14.

van Dijk, Jan. 2012. *The Network Society*. 3rd ed. London: Sage.

Van Grove, Jennifer. 2011. "Foursquare & Groupon Hook Up for Real-Time Deals." *Mashable*, July 29. http://mashable.com/2011/07/29/foursquare-groupon-partnership/

Van Grove, Jennifer. 2013. "Foursquare Checks In to More Revenue with Credit Card Specials." *CNet*, February 25. http://news.cnet.com/8301-1023_3-57571202-93/foursquare-checks-in-to-more-revenue-with-credit-card-specials/

van Zoonen, Liesbet. 2013. "From Identity to Identification: Fixating the Fragmented Self." *Media Culture & Society* 35 (1): 44–51.

Vaughan-Nichols, Steven J. 2011. "How Google—and Everyone Else—Gets Wi-Fi Location Data." *ZDNet*, November 16. https://www.zdnet.com/article/how-google-and-everyone-else-gets-wi-fi-location-data/

Vertesi, Janet. 2014. "Seamful Spaces: Heterogeneous Infrastructures in Interaction." *Science, Technology & Human Value* 39 (2): 264–284.

Waite, Emily. 2018. "How Maps Became the New Search Box." *Wired*, June 13. https://www.wired.com/story/how-maps-became-the-new-search-box/

Wakabayashi, Daisuke, and Douglas MacMillan. 2013. "Apple Taps into Twitter, Buying Social Analytics Firm Topsy." *Wall Street Journal*, December 2. http://www.wsj.com/articles/SB10001424052702304854804579234450633315742

Wang, Georgette, ed. 2013. *De-Westernizing Communication Research: Altering Questions and Changing Frameworks*. London: Routledge.

Warde, Alan. 1994. "Consumption, Identity-Formation and Uncertainty." *Sociology* 28 (4): 877–898.

Wark, McKenzie. 1994. *Virtual Geography: Living with Global Media Events*. Bloomington: Indiana University Press.

Wark, McKenzie. 2002. *Dispositions*. Applecross, Western Australia: Salt.

Warren, Samuel D., and Louis D. Brandeis. 1890. "The Right to Privacy." *Harvard Law Review* 4, no. 5 (December 15): 193–220.

Wasko, Janet. 2004. "The Political Economy of Communications." In *The SAGE Handbook of Media Studies*, edited by John D. H. Downing, Denis McQuail, Philip Schlesinger, and Ellen A. Wartella, 309–329. Thousand Oaks, CA: SAGE.

Wasser, Frederick. 2001. *Vini Vidi Vici: The Hollywood Empire and the VCR*. Austin: University of Texas Press.

Waters, Susan, and James Ackerman. 2011. "Exploring Privacy Management on Facebook: Motivations and Perceived Consequences of Voluntary Disclosure." *Journal of Computer-Mediated Communication* 17 (1): 101–115.

Wauters, Robin. 2015. "Apple Is Quietly Expanding Its Stealthy R&D Center in Lund (Sweden) to Work on 'Advanced Mapping Technology': Report." *Tech.eu*, September 15. http://tech.eu/brief/apple-mapping-technology-lund-sweden/

Weintraub, Seth. 2009. "Apple Purchased Placebase in July to Replace Google Maps?" *Computerworld*, October 1. http://www.computerworld.com/article/2467794/smartphones/apple-purchased-placebase-in-july-to-replace-google-maps-.html

Welch, Chris. 2013. "Foursquare Rolls Out 'Super-Specific' Filters to Help You Search Out the Perfect Destination." *The Verge*, May 22. https://www.theverge.com/2013/5/22/4355862/foursquare-rolls-out-super-specific-natural-language-search

Weld, Kirsten. 2014. "Washington's Prying Eyes." *NACLA Report on the Americas* 47 (4): 37–39.

Wesch, Michael. 2009. "YouTube and You: Experiences of Self-Awareness in the Context Collapse of the Recording Webcam." *Explorations in Media Ecology* 8 (2): 19–34.

Westin, Alan F. 1967. *Privacy and Freedom*. New York: Atheneum.

Whittaker, Zack. 2017. "Despite Privacy Outrage, AccuWeather Still Shares Precise Location Data with Ad Firms." *ZDNet*, August 25. http://www.zdnet.com/google-amp/article/accuweather-still-shares-precise-location-with-advertisers-tests-reveal/

Wiggers, Heiko. 2011. "Mind Your Privacy: American Tech Companies and Germany's Privacy Laws." *International Journal of Humanities and Social Science* 1, no. 11 (August): 210–220.

Wikipedia. 2017a. "Locative Media." *Wikipedia.org*. https://en.wikipedia.org/wiki/Locative_media

Wikipedia. 2017b. "Wardriving." *Wikipedia.org*. https://en.wikipedia.org/wiki/Wardriving

Wilken, Rowan, ed. 2012a. "Locative Media," special themed issue of *Convergence: The International Journal of Research into New Media Technologies* 18 (3). http://journals.sagepub.com/toc/cona/18/3

Wilken, Rowan. 2012b. "Locative Media: From Specialised Preoccupation to Mainstream Fascination." *Convergence: The International Journal of Research into New Media Technologies* 18 (3): 243–247.

Wilken, Rowan. 2014a. "Places Nearby: Facebook as a Location-Based Social Media Platform." *New Media & Society* 16 (7): 1087–1103.

Wilken, Rowan. 2014b. "Proximity and Alienation: Narratives of City, Self, and Other in the Locative Games of Blast Theory." In *The Mobile Self: Narrative Practices with Locative Technologies*, edited by Jason Farman, 175–191. New York: Routledge.

Wilken, Rowan. 2014c. "Twitter and Geographical Location." In *Twitter and Society*, edited by Katrin Weller, Axel Bruns, Jean Burgess, Merja Mahrt and Cornelius Puschmann, 155–167. New York: Peter Lang.

Wilken, Rowan. 2015. "Mobile Media and Ecologies of Location." *Communication, Research and Practice* 1 (1): 42–57.

Wilken, Rowan. 2016. "The De-gamification of Foursquare?" In *Social, Casual and Mobile Games: The Changing Gaming Landscape*, edited by Tama Leaver and Michele Willson, 179–192. New York: Bloomsbury.

Wilken, Rowan. 2017. "The Quick Brown Fox Jumps over the Lazy Dog: Perec, Description and the Scene of Everyday Computer Use." In *The Afterlives of Georges Perec*, edited by Rowan Wilken and Justin Clemens, 226–242. Edinburgh: Edinburgh University Press.

Wilken, Rowan. 2018a. "Social Media App Economies." In *The SAGE Handbook of Social Media*, edited by Jean Burgess, Thomas Poell, and Alice Marwick, 279–296. London: SAGE.

Wilken, Rowan. 2018b. "The Necessity of Geomedia: Understanding the Significance of Location-Based Services and Data-Driven Platforms." In *Geomedia Studies: Spaces and Mobilities in Mediatized Worlds*, edited by Karin Fast, André Jansson, Johan Lindell, Linda Ryan Bengtsson, and Mekonnen Tesfahuney, 21–40. New York: Routledge.

Wilken, Rowan, and Peter Bayliss. 2015. "Locating Foursquare: The Political Economics of Mobile Social Software." In *Locative Media*, edited by Rowan Wilken and Gerard Goggin, 177–192. New York: Routledge.

Wilken, Rowan, and Gerard Goggin. 2012a. "Mobilizing Place: Conceptual Currents and Controversies." In *Mobile Technology and Place*, edited by Rowan Wilken and Gerard Goggin, 3–25. New York: Routledge.

Wilken, Rowan, and Gerard Goggin, eds. 2012b. *Mobile Technology and Place*. New York: Routledge.

Wilken, Rowan, and Gerard Goggin. 2015. "Locative Media—Definitions, Histories, Theories." In *Locative Media*, edited by Rowan Wilken and Gerard Goggin, 1–19. New York: Routledge.

Wilken, Rowan, Gerard Goggin, and Heather Horst, eds. 2019. *Location Technologies in International Context*. London: Routledge.

Wilken, Rowan, and Anthony McCosker. 2012a. "The Everyday Work of Lists." *M/C: A Journal of Media and Culture* 15 (5). http://journal.media-culture.org.au/index.php/mcjournal/article/view/554

Wilken, Rowan, and Anthony McCosker. 2012b. "List: Editorial." *M/C: A Journal of Media and Culture* 15 (5). http://journal.media-culture.org.au/index.php/mcjournal/article/view/581

Wilken, Rowan, and Julian Thomas. 2019. "Maps and the Autonomous Vehicle as a Communication Platform." *International Journal of Communication* 13. https://ijoc.org/index.php/ijoc/article/view/8450

Wilkinson, Kenton T., and Patrick F. Merle. 2013. "The Merits and Challenges of Using Business Press and Trade Journal Reports in Academic Research on Media Industries." *Communication, Culture & Critique* 6 (3): 415–431.

Willems, Wendy, and Winston Mano. 2017. "Decolonizing and Provincializing Audience and Internet Studies: Contextual Approaches from African Vantage Points." In *Everyday Media Culture in Africa: Audiences and Users*, edited by Wendy Willems and Winston Mano, 1–26. New York: Routledge.

Williams, James. 2018. *Stand Out of Our Light: Freedom and Resistance in the Attention Economy*. Cambridge: Cambridge University Press.

Wilson, Matthew W. 2016. "Critical GIS." In *Key Methods in Geography*, edited by Nicholas Clifford, Meghan Cope, Thomas W. Gillespie, and Shuan French, 285–301. London: SAGE.

Wilson, Matthew W. 2017. *New Lines: Critical GIS and the Trouble of the Map*. Minneapolis: University of Minnesota Press.

Wingfield, Nick. 2012. "Apple Rejects App Tracking Drone Strikes." *New York Times*, August 30. http://bits.blogs.nytimes.com/2012/08/30/apple-rejects-app-tracking-drone-strikes/

Winseck, Dwayne. 2011. "The Political Economics of Media and the Transformation of the Global Media Industries." In *The Political Economics of the Media: The Transformation of the Global Media Industries*, edited by Dwayne Winseck and Dal Yong Jin, 3–48. London: Bloomsbury.

Wittel, Andreas. 2011. "Qualities of Sharing and Their Transformations in the Digital Age." *International Review of Information Ethics* 15 (9): 3–8.

Wong, Andrew. 2010. "Angel Finance: The Other Venture Capital." In *Venture Capital Investment: Strategies, Structures, and Policies*, edited by Douglas Cumming, 71–110. Hoboken, NJ: John Wiley & Sons.

Wood, David Murakami. 2017. "Spatial Profiling, Sorting and Prediction." In *Understanding Spatial Media*, edited by Rob Kitchin, Tracey P. Lauriault, and Matthew W. Wilson, 225–234. London: SAGE.

Wood, Denis. 1992. *The Power of Maps*. New York: Guilford Press.

Yeung, Ken. 2017. "Foursquare Launches Pilgrim SDK to Open Up Access to Location Intelligence in Apps." *VentureBeat*, March 1. https://venturebeat.com/2017/03/01/foursquare-launches-pilgrim-sdk-to-open-up-access-to-location-intelligence-in-apps/

Yoshida, Alex. 2014. "Two Laws Regarding Cellphone in Japan: No. 1 'Buying' a Local SIM." *BlogFromAmerica (English): Information on Japanese Cellphone/Data Access/WiFi/Others for Foreign Travellers*, June 23. http://eng.blogfromamerica.com/archives/43

Yoshida, Alex. 2014. "Two Laws Regarding Cellphone in Japan: No. 2 Requirement of Giteki Mark (Certification of Conformance to Technical Standards) under the Radio Act." *BlogFromAmerica (English): Information on Japanese Cellphone/Data Access/WiFi/Others for Foreign Travellers*, June 25. http://eng.blogfromamerica.com/archives/60

Yoshida, Alex. 2016. "Effective May 21st, 2016, It Is Legal for a Non-Japanese Foreigner to Use Your Non-Japanese Phone in Japan for Less than 90-Day at a Time." *BlogFromAmerica (English): Information on Japanese Cellphone/ Data Access/WiFi/Others for Foreign Travellers*, December 26. http:// eng.blogfromamerica.com/archives/116

Zahradnik, Fred. 2017a. "Assisted GPS, A-GPS, AGPS." *Lifewire*, June 7. https:// www.lifewire.com/assisted-gps-1683306

Zahradnik, Fred. 2017b. "How GPS Works on the iPhone." *Lifewire*, June 8. https:// www.lifewire.com/iphone-gps-set-up-1683393

Zeffiro, Andrea. 2012. "A Location of One's Own: A Genealogy of Locative Media." *Convergence: The International Journal of Research into New Media Technologies* 18 (3): 249–266.

Zeffiro, Andrea. 2015. "Locative Praxis: Transborder Poetics and Activist Potentials of Experimental Locative Media." In *Locative Media*, edited by Rowan Wilken and Gerard Goggin, 66–80. New York: Routledge.

Zhao, Bo. 2015. "Detecting Location Spoofing in Social Media: Investigations of an Emerging Issue in Geospatial Big Data." Dissertation Presented in Partial Fulfillment of the Requirements for the Degree Doctor of Philosophy in the Graduate School of The Ohio State University, Graduate Program, Ohio State University.

Zickuhr, Kathryn. 2012. "Three-Quarters of Smartphone Owners Use Location-Based Services." *PewResearchCenter*, May 11. http://www.pewinternet.org/2012/05/ 11/three-quarters-of-smartphone-owners-use-location-based-services/

Zickuhr, Kathryn. 2013. "Location-Based Services." *PewResearchCenter*, September 12. http://www.pewinternet.org/2013/09/12/location-based-services/

Zielinski, Siegfried. 2006. *Deep Time of the Media: Toward an Archaeology of Hearing and Seeing by Technical Means*. Cambridge, MA: MIT Press.

Ziman, John. 1980. *Teaching and Learning about Science and Society*. Cambridge: Cambridge University Press.

Zook, Matthew J. 2002. "Grounded Capital: Venture Financing and the Geography of the Internet Industry, 1994–2000." *Journal of Economic Geography* 2 (2): 151–177.

INDEX

Figures are indicated by *f* following the page number

Apple smartphone
 location capabilities of, xiv–xv
Apple Watch, 59
Application Programming Interface
  (API), 16, 83
"applications" sector
 in mobile location ecosystem, 31f, 33
App utility
 compartmentalization and, 123–127
AR (augmented reality), 220
Arcadia, 115, 115f, 116
ArcGIS software products, 48
art(s)
 introduction, 93–97, 94f
 redistribution of the sensible and,
    93–97, 94f
Arvidsson, A., 151
ASICS, 123
Assisted GPS (AGPS), xv
 benefits of, xv
AT&T, 177
"attention economy," 212
Attribution, 74
augmented reality (AR), 220
Australian Privacy Act, 167
"authentic self"
 performance of, 152
Auto Channel, 25
"automatically saved traces," 72f
Avast Group
 Location Labs of, 32

Bachelard, G., 141
Baleedi, J., 108
Bamford, J., 170, 182
Banjo, 12, 28, 78
Bann, S., 95
Barbour, K., 130
Barouch, J., 74
Barreneche, C., 84
Base Transceiver Station (BTS), 110, 118
Baudrillard, J., 102, 104
Baym, N., 77
Beck, U., 144
Begley, J., 95, 97, 102–109, 106f, 180, 216
Bellanova, R., 208–209
Berger, J., xiii, xiv
Best, J., 14
Betamax, 121
Bhaskar, M., 120

Big Data surveillance, 185
Bigo, D., 183
Bing Maps, 32
biography
 chosen, 144
 one's, 144
 "reflective," 144
 "reflexive," 144
Blank, S., 66
Blast Theory, 95, 97–102, 98f, 216
bloggers
 video, 128
Bookings.com, 59
Border Bumping, 95, 97, 109–115,
    111f–113f, 216
"BoundlessInformant," 179
Bowker, G.C., 40, 119
boyd, d., 128–129, 198
Brandeis, L.D., 190
Braun, J., 43
Bridle, J., 107
Brightest Flashlight, 158, 194
Brightkite, 12, 75, 195, 219
Brin, S., 49
BroadMap, 57–58
BT, 177
BTS. *See* Base Transceiver Station (BTS)
Bucher, T., 85, 109
Burdon, M., 85, 210
Bureau of Investigative Journalism,
    106–107, 180–181
Burns, K., 195
Burstly ad library, 158
Bush, G.W., Pres., 103, 186
Butler, J., 144
Button, 29, 42
Bygrave, L., 212

Cable, J., 182
Cadenasso, M., 27
Carnegie Mellon University, 158
Cartographic Services, 47
Caspar, J., 161
cataloguing
 described, 139–140
 on Foursquare, 139–142
Catch, 58
CDMA (Code Division Multiple Access)
    wireless technology
 of Qualcomm, 26